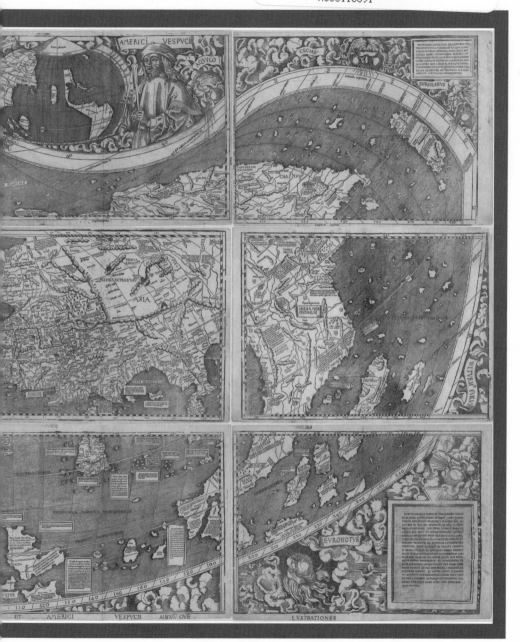

MORE PRAISE FOR *PUTTING "AMERICA" ON THE MAP*

"Adhering to the highest standards of scholarly thesis, Dr. Schwartz has taken a complex subject and created an accessible narrative that has broad appeal. The story . . . is brilliantly told through the voice of a passionate collector and historian."

—Margaret Beck Pritchard, curator of prints and maps for the
Colonial Williamsburg Foundation, and coauthor of
Degrees of Latitude: Mapping Colonial America

"Based on extensive archival and cartobibliographic research, Schwartz weaves an amazingly extensive and comprehensive story about one map—Martin Waldseemüller's 1507 world map, recognized as the most important cartographic document in the history of the Americas. He traces the map's five-hundred-year history from its creation, placing it in the context of the contemporary political and intellectual environment of the European period of discoveries, to the recent acquisition of the only surviving copy by the Library of Congress, a project that was spearheaded by the library's German-Dutch specialist Margrit Krewson and the Geography and Map Division. Since this excellent historiographic study brings together the known primary and secondary literature relevant to the history of the map and discusses the controversies that have arisen, it will be the beginning point for future research on this fascinating topic."

—Ronald E. Grim, curator of maps for the
Norman B. Leventhal Map Center at the Boston Public Library

Putting
"AMERICA"
ON THE MAP

Martin Waldseemüller. Courtesy of Collection
Médiathèque Victor Hugo, Saint Dié-des-Vosges.

SEYMOUR I. SCHWARTZ

Putting
"AMERICA"
ON THE MAP

THE STORY OF THE MOST IMPORTANT GRAPHIC
DOCUMENT IN THE HISTORY OF THE UNITED STATES

 Prometheus Books

59 John Glenn Drive
Amherst, New York 14228–2119

Published 2007 by Prometheus Books

Inquiries should be addressed to
Prometheus Books
59 John Glenn Drive
Amherst, New York 14228–2119
VOICE: 716–691–0133, ext. 210
FAX: 716–691–0137
WWW.PROMETHEUSBOOKS.COM

11 10 09 08 07   5 4 3 2 1

Library of Congress Cataloging-in-Publication Data

Schwartz, Seymour I., 1928–
Putting "America" on the map : the story of the most important graphic document in
  the history of the United States / Seymour I. Schwartz.
      p. cm.
Includes bibliographical references and index.
ISBN 978–1–59102–513–9
  1. America—Name. 2. Waldseemüller, Martin, 1470–1521? 3. America—
Maps—Early works to 1800. 4. America—Historical geography. 5. America—
Discovery and exploration—Maps. 6. World maps—History. 7. Cartography—
History. I. Title.

E18.75.S39 2007
973—dc22
                                                                          2007022204

Printed in the United States of America on acid-free paper

CONTENTS

6 CONTENTS

FOREWORD

Dr. Seymour Schwartz, an emeritus distinguished alumni professor of surgery, has asked me to write a preface to his carefully researched and highly informative history of the 1507 Waldseemüller map of the world and its acquisition by the Library of Congress. Because the genesis and announcement of the Waldseemüller map took place five hundred years ago, in late April 1507 (although there is good evidence that the actual copy now in the library was probably printed in Strasbourg in 1516), Dr. Schwartz's account is apposite and timely. It will become an essential companion to the map when the library puts it on display, but the book stands on its own, as it details an important aspect of America's past.

Dr. Schwartz dissects the controversies that have surrounded many aspects of the Waldseemüller map since its accidental discovery in 1901 by Father Joseph Fischer in Wolfegg Castle, Germany. Questions unanswered—and in some cases unanswerable—are set out; and the claims of ardent proponents of various theories about the map's sources, priority, and accuracy are fairly summarized. In addition to detailed descriptions (which are substantially complemented by illustrations) of the many maps involved in this complex story, Dr.

Schwartz explores the availability of geographical and cosmological information at the beginning of the sixteenth century. We cannot be sure, of course, how much of what could have been known actually was known by the members of the Gymnasium Vosagense in St. Dié who publicized in April 1507 the forthcoming appearance of the map.

Certain details of the Waldseemüller map are so unexpectedly accurate—the western coast of the southern tip of South America below fifty degrees south latitude, the clear indication of a western ocean between the American continents and Asia, the shape and mountainous nature of what is now the western coast of Chile, the delineation of southern stars—that they suggest some (probably Portuguese) mariners must have navigated through the straits of the islands and around Cape Horn before Magellan. But no one knows; there is no evidence, only surmise.

There are other mysteries and near-detective stories related by Dr. Schwartz. Not least, the seemingly improbable export from Germany of a map that was on the select list of protected German cultural property. The two people who sustained the effort and ultimately secured the exit permit were Prince Johannes Waldburg-Wolfegg, the owner of the map, who agreed to sell it, and Margrit Krewson, the former specialist of the German-Dutch section of the Library of Congress, whose persistence and persuasiveness with high-ranking members of the German government and industry actually led to the final positive decision. Mrs. Krewson considers the acquisition of the map her legacy to the Library of Congress and the American people.

Dr. Schwartz also describes in some detail the history of the Library of Congress, including the nature of its relevant German and Geography and Map Division holdings. These are but two of the many collections that make the Library of Congress the largest library in the world. The Waldseemüller map will be displayed in the Thomas Jefferson Building, the library's oldest building, appropriately set in the second-floor Hall of Discovery.

It is, I suspect, the connection with the Jefferson Building that induced Dr. Schwartz to ask me to write a brief preface. As director of scholarly programs at the Library of Congress for fifteen years, I worked in the Jefferson Building and am familiar with the library's structure and some of its expert curators and specialists, which give the Waldseemüller map a fitting context. Also, as a historian, I am aware of the lacunae and ambiguities, the multiple possibilities and loose ends inherent in any attempt to reconstruct complex events of five or six centuries ago. Dr. Schwartz has done an admirable job of putting before us the evidence and the controversies about the naming of America. All of us must now navigate these storm-tossed waters before landing on the terra firma of our own conclusions.

Prosser Gifford
Former Director of Scholarly Activities
for the Library of Congress
Woods Hole, Massachusetts
September 2006

PREFACE

It is a privilege to comment on the cartography of the unique 1507 twelve-sheet, 1.3- × 2.34-meter woodcut wall map of the world by Martin Waldseemüller. Since its discovery in 1901, scholars have recognized the supreme significance of this monumental achievement of early European cartography. It is the first printed map of the world to delineate the concept of the existence of a "New World" based on the news of the earliest recorded transatlantic explorations by Europeans.

My longtime good friend, client, colleague, medical adviser—and internationally renowned surgeon—Dr. Seymour I. Schwartz, describes here in great detail the early twentieth-century discovery and publication of this highly important, unique historical artifact. He discusses its origins, physical properties, provenance, and impact on scholarship. Dr. Schwartz presents a scientific and social history of the map, including an explanation of why Waldseemüller elected to name the "New World" "America" rather than "Columbia."

My observations on Waldseemüller's cartography, since those expressed in the *Atlas of Columbus and the Great Discoveries* in 1990, relate mostly to speculation about numerous unanswered queries still surrounding this historic document. *Putting "America" on the Map* relates the saga of the relocation of the map and its associate documents, from their centuries-long residence in a German castle to a permanent home at the Library of Congress. Having moved from private to public possession, this Renaissance treasure has become a focal point of renewed interest, allowing researchers the opportunity to investigate many of the map's perplexing questions and to solve some of its many mysteries.

Martin Waldseemüller, a German scholar-priest, gifted writer, mapmaker, and printer, settled at the court of Duke René II of Lorraine at St. Dié early in the sixteenth century, where he joined a small group of humanists known as the Gymnasium Vosagense. One of the most dramatic achievements of this group was the compilation and production of Waldseemüller's cartographic masterpiece, the *Universalis Cosmographia* of 1507. That the mapmaker was a transitional figure from the late medieval to early Renaissance thought is manifest in his employment of Ptolemaic delineations for much of Europe and Asia, while incorporating newly received information from contemporary explorers.

A succession of fifteenth-century Portuguese navigators progressively penetrated the unknown west coast of Africa, eliciting novel geographical concepts. Bartolomeu Dias rounded the Cape of Good Hope in 1487–88. By 1499, Vasco da Gama had sailed across the Indian Ocean, becoming the first European to navigate the sea route to India. Cartographers then completed the continent of Africa with remarkable accuracy, and the "inland sea" concept of the Indian Ocean inherited from Ptolemaic maps of the world was dispelled, as shown on Waldseemüller's map. To the west, information came via the four voyages of Columbus and other explorers sailing for Spain, and also from the controversial Amerigo Vespucci, who claimed to have made two Atlantic crossings as a pilot on Spanish ships and two others under the Portuguese flag. Columbus was sailing westward to discover a shorter, more direct route to the Far East than the Portuguese-controlled route around Africa when he ran into the Caribbean Islands. He believed he had reached Asia. Because of the unexpected appearance of what was later called the Western Hemisphere—North America, Central America, and South America—much of the earliest information reaching European mapmakers was speculative, inaccurate, and often contradictory.

Three manuscript charts have survived that are important to answering questions surrounding the first fifteen years of the European discovery of America—and of modern speculation about Waldseemüller's sources. The first, drawn from personal experience by one of Columbus's pilots, Juan de la Cosa, and signed and dated 1500 (possibly completed as late as 1510), is the only surviving chart of Columbus's first voyages. La Cosa combined on different scales traditional delineations of the Old World with his dramatic first presentation of the New World. He added indefinite coastlines representing northeastern South America, and to the north, suggestions of East Asia that would evolve into being North America. Because of its Spanish origin, this delineation no doubt joined other charts of discovery carefully secreted at Seville.

The second chart, known as the "Cantino Planisphere" of 1502, was copied from the equally secret official Portuguese master chart at Lisbon. Albert Cantino, an agent of Ercole d'Este, Duke of Ferrara, acquired it surreptitiously for the duke in 1502. It celebrates the enormous Portuguese accomplishments in the east, around the Cape of Good Hope, across the Indian Ocean to India and Southeast Asia. For the west, the New World, the mapmaker presumably had access to Spanish sources, as Cuba and Hispaniola are clearly drawn. The enigmatic landmass northwest of Cuba was either an attempt at showing East or Southeast Asia, as Columbus believed, or—according to the eternally posited theory—that it represented southeastern North America and the Florida peninsula (which it strongly resembles).

Cantino was the first to delineate the meridian line proclaimed at the Treaty of Tordesillas, signed by Spain and Portugal in 1494, which divided newly discovered territories. Portugal was to receive lands east of the line, and Spain those to the west. In the north a large island appeared, divided by the demarcation line. Cantino placed a Portuguese flag and declared "Land of the King of Portugal" on the eastern part, reflecting claims of discovery of northeastern Canada by Gaspar Corte-Real in 1501. The two divisions are delineated differently; the eastern sector is carefully detailed topographically, while the west is traditionalized, implying the possibility of the two sections coming from different sources. In the south, the demarcation line bisects the large landmass beginning just north of the equator, thereby creating Portuguese Brazil.

The third of these great nautical planispheres was drawn by Nicolo Caveri of Genoa sometime between 1504 and 1505. Caveri's circles of compass roses, his distance scales, and particularly his innovative presentation of a latitude scale exhibit considerable progress in the art and science of cartography at the beginning of the sixteenth century. For the New World in general, he must have had access to the Cantino map or a common source, and for the Brazilian coast, probably reports of the Portuguese voyages of 1501–1504. Caveri had to have used a different source than Cantino for the Arabian coast and the Red Sea, as they differ sharply.

Waldseemüller's magnificent woodcut world map employed numerous sources in addition to the second-century CE theories of Claudius Ptolemy; some can be traced, some not. The Frau Mauro map of ca. 1459, known in Lisbon, provided Marco Polo's concepts; and the larger Henricus Martellus manuscript of ca. 1489 had modified and improved aspects of Ptolemy's world in addition to indicating Portuguese discoveries in Africa and Asia. Martellus was the first to show that Dias had rounded the Cape of Good Hope, and, importantly, his map displayed the division of latitude *and* longitude into degrees. By combining classical, medieval, and contemporary sources, Waldseemüller created the first published description of the world as its unknown parts were being discovered. It was the first printed version of the Lusitino-Germanic group of maps, and, since one thousand copies were printed, it influenced more people than its manuscript predecessors.

One mystery yet to be solved is the possibility that this map was employed to advance Portuguese claims regarding the definite location of the line of demarcation referred to above. Two unusual features appear in the northern ocean separating Europe from the New World. First, the northeast extremity of the new lands is terminated by a straight edge, while an island far to the east, possibly representing the Newfoundland-Labrador of Corte-Real, has its western edge terminated in the same straight-edge manner. This seems to imply that wherever the line fell in the North Atlantic, Canada should belong to Portugal.

In addition, farther south, off the West African coast, a large section of the woodblock appears to have been cut out and reworked, possibly to reposition the Cape Verde Islands farther west. These maneuvers could represent changes approved by the pope in 1506 that carved off the large triangle of South America, which became Portuguese Brazil. This then is the most current surviving map available regarding the newly discovered lands. It furthered Portuguese interests even while Waldseemüller did not attempt to portray the line itself.

Most baffling is the appearance of a striking if not detailed western coast of North America, Central America, and South America, implying knowledge of an ocean to the west, with a gap that if true would have permitted passage from Atlantic to Pacific approximately at the location of Panama. Furthermore, on the inset map at the upper right, this gap is eliminated by drawing the continents as joined. No less intriguing is the Portuguese flag, with its staff placed by Waldseemüller at the very southern tip of South America, appearing as if it were just south of a *strait*. These features imply that Waldseemüller may have seen a chart from an earlier Portuguese navigator than history has so far recorded.

Waldseemüller later drew from Caveri's delineation (described above) for his 1513 edition of Ptolemy's geography, to which he added important new maps signaling recent discoveries. In addition, Waldseemüller's great "Carta Marina" of 1516 is virtually an enlarged copy of Caveri's chart. Significantly, we still do not know what brought the Iberian information to the small intellectual center at St. Dié in the

Vosges Mountains. Was it the reputation of the gymnasium, or of Waldseemüller, or the influence of the Duke of Lorraine? We do know that the influence of the cosmographical activities in St. Dié, led by Waldseemüller, was profound.

Kenneth Nebenzahl
Glencoe, Illinois

ACKNOWLEDGMENT

Many individuals have had significant roles in the evolution of this work. My deceased wife, Ruth, started it all when she brought cartography into my life and encouraged my passion for the subject. The research was greatly facilitated by the hospitality extended by encouraging personnel at three pertinent locations. At the Ignatiushaus in Munich, archivist Dr. Rita Haub provided all of the original documents related to Father Joseph Fischer, SJ, the discoverer of the Waldseemüller world map of 1507. Dr. Bernd Mayer, curator of the Wolfegg Castle Collection, with the permission of Prince Waldburg-Wolfegg, conducted a tour of the castle and brought forth for study material concerning the map and the Dürer star chart. The extensive material, consisting of memoranda and correspondence, which extended over almost a century, provided by the Geography and Map Division of the Library of Congress, was collated through the efforts of Patricia van Ee and James Flatness. Their constant input of documents and encouragement served as a recurrent stimulus. Heather Wanser of the Library of Congress's conservation department provided valuable input. Arthur Dunkelman provided the material from the Jay Kislak Foundation.

Just as she was a persistent catalyst for the transfer of the map from Wolfegg Castle to the Library of Congress, Margrit B. Krewson remained an invaluable resource for this work. Not only did she provide her extensive personal file focusing on the negotiations with Prince Waldburg-Wolfegg and submit to interview, but she also expressed continued interest in the project during innumerable telephone conversations.

Christopher Hoolihan translated the Latin text. Brigitte von Kessel translated the unpublished biography of Father Fischer. The manuscript was read and constructively criticized by Melinda Beard and three of my medical colleagues, Drs. James Adams, Irwin Frank, and Marshall Lichtman. My agent, John Silbersack, suggested significant modifications regarding the presentation of the subject matter. A major participant, whose involvement extended over a year, was Joe Avitable, a graduate student in American history, who shepherded me through the acceptable protocols of presenting references and citations. Gianna Nixon's graphic art expertise expedited the production. The willingness of two distinguished authors, Lucia Perelli and Edward P. Jones, to allow reference to their writings is greatly appreciated. I also greatly appreciate the enthusiastic contributions of my editor, Linda Greenspan Regan.

Whilst my Physicians by their love
*   are growne*
Cosmographers, and I their Mapp,
*   who lie*
Flat on this bed.

                                        John Donne

INTRODUCTION

Journey over all the universe in a map, without the expense and fatigue of traveling, without the inconveniences of heat, cold, hunger and thirst.

Miguel de Cervantes
Don Quixote

J ust as Tristram Shandy can pinpoint the moment of his conception by the time his mother asked his father whether he had wound the grandfather clock, I can define with precision what was actually the distant origin of this work. It was January 1963; I was approaching my thirty-fifth birthday; I was an academic surgeon who had achieved sufficient success to have been granted tenure by my university. My life had been focused, with total immersion in the art and science of my profession. I could converse in depth and with enthusiasm on only one subject—surgery. At that time, my physician wife suggested that I should broaden my vistas, develop diverse interests, or, as she put it, get a hobby.

My immediate response was that I had little spare time, and, if she felt so strongly, that she should find a hobby for me. The next day, before returning home from work, she crossed the street that her office faced and entered a secondhand bookstore. Briefly scanning the stacks, she selected and purchased for fifty cents a reasonable copy of the Bonanza Books edition of *Maps and Map-Makers* by R. V. Tooley. It was one of several birthday gifts and was presented with the suggestion that the subject, because of its disparity with surgery and its relation to history, which had always appealed to me, might tweak my interest. It was the first time that the word *cartography* entered my vocabulary.

Several months later, during a trip to New York City, I entered the Argosy Book Store on 59th Street near Lexington Avenue, which someone had suggested that I visit. After riding an old elevator that rose leisurely under the guidance of a casually attired attendant, I arrived at the map and print section on the second floor, where I was greeted by Mrs. Cohen, the store owner's wife. After an hour of nondirected perusal of many maps, varying in age, size, and focus, I somewhat sheepishly purchased a small 1795 map of the state of New York because it depicted the state before Rochester, the city I lived in, was settled. The price was twenty-five dollars—a very modest beginning.

About six months later, while attending a surgical meeting in Philadelphia, I visited the famous Sessler's old book and map store and, after an instructive conversation with the proprietress, I purchased Henricus Hondius's 1656 derivative of John Smith's 1612 map of Virginia for sixty-five dollars. I was still dabbling. Once again, in the course of a surgical convention, my probing continued. This time my interest rose both figuratively and literally. A rapid elevator ride to the twenty-eighth floor of 333 N. Michigan Avenue in Chicago led to the wonders of the establishment of Kenneth Nebenzahl, who has since been dubbed the acknowledged "grandfather" of the American antiquarian map trade.

This was the beginning of a long-term association with a man who became a mentor and friend. Many maps and pertinent books found their way from his shop to my home over the years. My fate was sealed by my mentor in 1971, when he asked whether I planned to, as he put it, become a true collector of cartographic Americana, or if I would just continue to purchase an occasional map. With a lack of appreciation as to the implications of being a "true collector," I responded that I would like to develop a meaningful collection focusing on printed maps of North America published before 1800.

This declaration permitted Kenneth Nebenzahl to don his mantle of salesmanship and immediately suggest that I purchase an important map he had available for sale, namely, Johannes Ruysch's world map of 1508. This is the earliest printed map depicting the discoveries in the Western Hemisphere available to a collector. Although the price generated much anxiety, needless to say, I bought the map, and my plebeian probing began its evolution into a passion with a focus on maps of North America extending from the era of discovery, to the period of settlement and expansion, to the achievement of independence, and to the ultimate establishment of the United States of America—a period encompassing about three hundred years.

Each acquisition generated a romantic involvement between devotee and document. Several maps have evoked periodic intellectual genuflection. The quest for additions to the collection has had its holy grails, a small number of essentially unobtainable icons. The earliest manuscript to present a cartographic picture of both Columbus's and Cabot's discoveries in the New World, the Juan de la Cosa map of 1500, is proudly displayed in the Museo Naval in Madrid. Fortunately, over the past two decades, I have participated every third year in a surgical conference in that city. This afforded me the opportunity to reverently view the map on many occasions.

In the case of the focus of this book, Martin Waldseemüller's world map of 1507—the document that established America's name—my involvement has been even greater. The first episode was a personal pilgrimage to the National Museum of American History of the Smithsonian Institution in Washington, DC, in 1983, at which time the folio containing the map was on display, opened to the sheet that shows the word "America" on the southern continent of the New World. Subsequently, in 1992, as part of the quincentennial celebration of

Columbus's discovery, the map's twelve joined sheets were presented in the National Gallery of Art, Smithsonian Institution, thus displaying the map for the first time in its entirety as a unit. I stood there viewing the monumental work, fixed in awe for well over an hour.

Imagine my reaction when I recently learned that there was a distinct possibility that the map's place of permanent residence might become our nation's Library of Congress. As a deeply interested citizen with a passion for maps that chronicle the history of this country, I contributed, in a small way but with a loud voice, in soliciting members of Congress to appropriate the funds required for acquisition. And, after the purchase had been achieved, my excitement was amplified when, on the evening of July 23, 2003, I witnessed the map's unveiling at the Library of Congress. The occasion offered me the opportunity to express appreciation to Prince Johannes Waldburg-Wolfegg of Germany for his willingness to part with the treasure in deference to what he regarded as a more appropriate home. America's elegantly graphic certificate of birth would forever reside on the land it had named.

Following the precept of sixteenth-century philosopher Sir Francis Bacon, that "reading maketh a ready man, conversation a knowing man, and writing a precise man," I decided to write a detailed history of a document that is regarded as one of the most important printed items related to this nation. In preparing for a definitive book, it was deemed obligatory to go to the primary sources, when possible, rather than rely on the words of others. This led to a flight across the same ocean that the map had recently taken, but in the opposite direction, to Europe. Visits marked by excitement and satisfaction were made to places with names that recur throughout the narrative: Munich, Saint Dié, the Vosges Mountains, Freiburg im Breisgau, Konstanz, and Wolfegg.

In Munich I was able to visit the Bavarian State Library, where my hands, covered by a pair of white cotton gloves, handled with care the 1486 edition of Ptolemy that was published at Ulm and contained, as a later insertion, one of the four known copies of Martin

Waldseemüller's globe gores (tapering segments of a map that can be affixed to a sphere to make a globe). Just one block away and across Ludwigstrasse on the second floor of the Universitätsbibliothek, I turned the pages of a copy of the *Cosmographiae Introductio*, published at St. Dié in 1507 into which Henricus Glareanus had inserted, sometime around 1510, two manuscript maps, made with pen and ink and colored, that he had, by his own assertion, copied from Martin Waldseemüller's world map. On both manuscript maps, the word "America" was inscribed.

Also in Munich a day was spent at the Ignatiushaus of the Society of Jesus, where Father Joseph Fischer, who fortuitously discovered the world map of 1507 at the Wolfegg Castle in 1901, worked from 1939 to 1941 and where most of his papers are archived. Dr. Rita Haub, the curator, graciously made this material, which was stored in seven large cardboard boxes, available. Revealing correspondence and meaningful notes spewed forth from their unimpressive containers. Munich provided material of matchless importance.

A five-hour drive south through Freiburg im Breisgau and then across the Rhine into France led to the foot of the Vosges Mountains. Following the famous Vosges wine trail brought my journey of primary research to the city of Saint Dié-des-Vosges, where the map and its accompanying narrative were conceived and executed, the location of the map's genesis. Arrival at the modern city after a tiring drive evoked an immediate feeling of disappointment, in that there was no sense of history immediately apparent. But the next morning, accompanied by the chimes of the cathedral, the atmosphere was transformed. The twelfth-century cloister and fifteenth-century section of the church brought me back to the time of interest. The Médiathèque Victor Hugo Library contained a treasure trove of significant documents and images.

Leaving Saint Dié-des-Vosges and returning along the same wine trail that brought me there, I crossed the Rhine into Germany and continued on to Freiburg im Breisgau. I had hoped to visit the regional university where Martin Waldseemüller, the cartographer, and Johann Schött, the printer—both of whom played nuclear roles in the ensuing history of America's baptismal document—were educated. Unfortunately, no fifteenth- or sixteenth-century building had withstood onslaughts of time and wars. After a brief visit to the famous Freiburg church, known as the Münster, the journey proceeded to the city of Konstanz, where Martin Waldseemüller, the man credited with authorship of the 1507 world map, had been designated a canon. Along the way, bold black letters on yellow road signs indicated the turn-offs to Radolfzell, which some biographers consider to be Waldseemüller's birthplace, and to Feldkirch, where Father Joseph Fischer was teaching geography in 1901 when he discovered the map that named America at Schloss Wolfegg (Wolfegg Castle).

It was at Wolfegg Castle where ultimate gratification of my modern expedition of discovery was achieved. Dr. Bernd Mayer, curator of the collection at the castle, conducted a tour of its halls, but, more important, provided historical descriptions, details, and data that are essential to the presentation of an exact and complete history of the birth document, which merits the adjectives *seminal*, *unmatched in importance*, and *revered*.

And now that Martin Waldseemüller's world map of 1507, the most expensive acquisition in the history of the Library of Congress, has assumed its appropriate position as the premier cartographic document in that edifice, it is timely, as we have reached the quincentennial of the map's production, to detail its history and to address the many intrigues and mysteries related to the document itself and the world it depicts.

CHAPTER 1
GENESIS
THE PLACE AND THE PARTICIPANTS

The invention of printing and the Reformation are and remain the two outstanding services of central Europe to the cause of humanity.

Thomas Mann

The Waldseemüller world map of 1507 is arguably the most significant printed graphic document among the many treasures of the United States of America. The size and artistry of its twelve sheets when joined constitute an approximately thirty-four-square-foot document. Its grandeur appropriately complements the map's historic presentation to the world of a name—"America"—that was incorporated by thirteen states on the northern continent in the Western Hemisphere at the time they united to create a new nation.

The locale where, during the Renaissance, the Christian world first bestowed a Christian name on a recently discovered continental mass in the New World could hardly have been anticipated. The momentous event took place neither in the cathedral of the contemporary Spanish monarchs—King Ferdinand II and Queen Isabella (who died in 1504 before the naming occurred), the Catholic rulers

27

of the nation credited with the discovery—nor in the seat of Christianity, at St. Peter's Cathedral in Rome, under the aegis of the contemporary Pope Julius II. Either of these venues would have been in keeping with the importance and impact of the event. Rather, the "baptismal ceremony" that used printer's ink and a press instead of holy water and a priest's anointing finger played out in an obscure locale specifically dedicated to the learning and dissemination of secular knowledge. This activity was distinct from ecclesiastic control. The event relied upon the process of printing that employed movable type, which was introduced only about fifty years earlier. The essence was the assignment of the name "America" to land in the Western Hemisphere on April 25, 1507, in the town of Saint Dié, in what was then designated as the Duchy of Lorraine.

As the twenty-first century unfolds, the city of Freiburg im Breisgau in southwestern Germany, about forty miles from Saint Dié, is a reasonable point of reference, both geographically and historically, for the beginning of a modern journey of scholastic pursuit. Freiburg im Breisgau is the location of the university where both Martin Waldseemüller, credited as the author of the extraordinary world map of 1507 that named America, and Johann Schött, who printed the 1513 edition of Ptolemy's *Geographia*, were educated in the latter half of the fifteenth century. That edition of Ptolemy's *Geographia* was edited by Martin Waldseemüller and plays a significant role in the analysis of the world map of 1507.

In Freiburg im Breisgau, what has been referred to as the most magnificent spire in all of Europe still stands. It identifies the cathedral known as the Münster, the construction of which began about 1200 and was not finished until 1513, the same year that the edition of the Ptolemy atlas containing significant maps by Martin Waldseemüller was published. Although the church suffered only minor damage during intensive bombing in World War II, the university buildings, where Waldseemüller and Schött studied, were destroyed.

A modern bridge that crosses the Rhine in the region of Freiburg im Breisgau provides entrance into the northeastern portion of France and the region now known as Alsace-Lorraine. The first city encountered on the French side of the river is Colmar, which houses, in the Musée d'Unterlinden, the famous Issenheim Altarpiece that was painted by Matthias Grünewald in 1512–16. Proceeding in a northwesterly direction along N 415, referred to as the Route de Vin des Vosges—passing Kaysersberg, the birthplace of Nobel Peace Prize winner Albert Schweitzer, and crossing the Meurthe River—brings the traveler to the modern city now named Saint-Dié-des-Vosges, with about 22,500 inhabitants known as "les Déodatiens."

The Vosges Mountains begin not far from Basel, Switzerland, and run in a northeastward direction in the northeastern part of France to the west of the Rhine River. The mountains contain rock salt, iron ore, and small amounts of gold and silver. The eastern slopes nurture the vines of the regional grapes, while the western side contains many streams that form the rivers flowing westward. Saint-Dié-des-Vosges is located at the base of the western slope, where the Robache Rapids enforce the Meurthe River. The city, about forty miles equidistant from Nancy and Strasbourg, is bordered by large forests and serves as the capital of the region known as Massif Vosgien. The name, Saint-Dié-des-Vosges, was first used by the postal service in the nineteenth century, but the use of that name was limited until the twentieth century. The designation that appears on most current maps is simply "St. Dié," and, since the community contains no significant hotel or restaurant, it is not mentioned in most modern books that provide information for travelers to France.[1]

Currently, the city memorializes its role in the naming of America with several plaques placed conspicuously, either in the pavement or alongside a walkway, around the cathedral that contains architectural elements of the twelfth through the eighteenth centuries. Adjacent to the cathedral, the Médiathèque Victor Hugo Library houses among its 180,000 items 600 manuscripts and 140 incunables (books printed before 1500). The highlight of the library is the La Salle du Trésor (Room of Treasures), presenting a depiction of the act of "baptism" of America at Saint Dié in 1507. One street in the modern city, rue d'Amérique, is intersected by rue Gymnase Vosgien, which honors the group of men whose combined scholarship contributed to the naming of America.

A consideration of the city's history, pertinent to its role in the naming of America, in a reverse chronological sequence most appropriately would begin in 1990. At that time, the International Festival of Geography was initiated as a continuation of the Gymnasium Vosagense's work that resulted in the naming of America. Currently, over thirty thousand people annually attend conferences dedicated to understanding the evolution of geographic knowledge. The attendees are greeted by a large sculpted globe that dominates one of the entrances into the city. The conference periodically honors a distinguished geographer with the Vautrin Lud International Award, taking its name from the man responsible for creating the Gymnasium Vosagense in the early sixteenth century.

Continuing in a reverse chronology, after World War II, the United States provided crucial assistance to the city recently destroyed by the Germans. In July 1911, Albert Lebrun, the French Minister of the Colonies, and Robert Bacon, the ambassador from the United States to France, were present in Saint Dié during the unveiling of a commemorative plaque marking the "baptismal house of America." That event stemmed from the singular efforts of the American journalist Charles Heinrich. Because the fourth centenary (1907) of the naming of America in 1507 was overlooked by both the city of Saint Dié and the United States, Heinrich created in New York City, in 1908, the St.-Dié Society to remember the event. He also started a journal, the *St.-Dié Press*, and he proposed that, because 1511 marked the death of Matthias Ringmann, an important member of the Gymnasium Vosagense, the fourth centenary of the naming of America should be celebrated in 1911.

Almost two decades earlier, at the Chicago Exposition that opened in 1893 to celebrate the fourth centenary of the discoveries of Christopher Columbus, an entire room was dedicated to the naming of America. The presentation included two examples of the *Cosmo-*

graphiae Introductio, the 1507 printed narrative that was produced at Saint Dié and that presented the word "America" for the first time with an explanation for the assignment of that name. One more relationship between Saint Dié and America is historically significant. Jules Ferry, who was born in Saint Dié in 1832, became president of the Council of Ministers and minister of Foreign Affairs after the fall of the Second Empire. As a consequence of his position, on July 4, 1884, he and L. P. Morton, United States Minister Plenipotentiary to France, signed the Deed of Donation, declaring a gift from France to the United States—the Statue of Liberty. It served as a commemorative monument celebrating liberty and the lasting union of France with the new nation on the eve of the centenary of its independence. On July 24, 1886, on Bedloes Island in New York Bay, the statue's sculptor, Auguste Bartholdi, unveiled the monument before a large crowd that included President Grover Cleveland and Ferdinand de Lesseps, the French diplomat-engineer who had designed the Suez Canal. Jules Ferry did not attend, because a year earlier the government with which he was associated had fallen and he no longer held office.[2] Thus, the same relatively small community had a role in both naming America and contributing one of the most significant icons of a nation that bears that name.

The genesis of Saint Dié—its evolution during the early sixteenth century and the role it played in the naming of America—takes us back to the Bronze Age. Tools of flint and bronze artifacts that have been uncovered provide evidence for the existence of Neolithic inhabitants in the region. The discovery of remnants of Gallo-Roman walls and imperial coins that date from the first through fourth century CE furnish evidence for the existence of a trading site for traveling merchants. These merchants supplied provisions for the local farmers and the miners who extracted iron ore from the mountains during the Roman occupation.[3]

The history of Saint Dié as a community dates from the middle of the seventh century with the arrival in the area of an evangelic bishop from Nevers, France, named Déodat, Dieudonné, or in the Latinized form, Deodatus. Deodatus had followed the road from Colmar to Deneuve and stopped at the left bank of the Meurthe River, where he had an epiphany and then built an oratory. Later, he chose a site on a butte on the right side of the river and created a humble cell in which he lived. At the foot of Mount Saint Martin, where the Robache River joins the Meurthe River, Deodatus built a monastery that he dedicated to Saint Benoît and called it the Jointures, because of its location at the juncture of two rivers. Below the monastery, farmers and their wives established a settlement consisting of thatched cottages and farmyards that provided subsistence while they worked under the direction of monks and expressed their religious devotion. Deodatus died in the village, which he had founded, on June 19, 679. His remains were initially preserved in the ancient oratory known as the chapel of Petit Saint-Dié, but, at some unknown later date, they were transferred to the collegiate church. The sarcophagus that contained his remains was destroyed when the building was dynamited during World War II. In addition to the purported miracles that occurred at the site of the sarcophagus, a number of purported miracles performed during his travels in Alsace-Lorraine and the Vosges Mountain region were attributed to Deodatus and eventually led to his canonization as Saint Deodatus, or Saint Dié.[4]

Within a century of the initial establishment of a settlement in the region of Saint Dié, the forests were pushed back to the mountains and more peasants arrived in what became known as the Valley of Galilee. In 769, Charlemagne (Charles I, or Charles the Great)—the King of the Franks from 768 to 814 and founder of the first empire in western Europe after the fall of Rome—made a gift of the monastery of Saint Dié to the Abbey of Saint Denis, which was run by his friend, Fulrad. Fulrad, who would eventually be canonized as Saint Fulrad after his death, was specifically charged with maintaining about a dozen monks at the monastery of Saint Dié to pray for the soul of the king's father.[5]

In the ninth century, Frederick, the Duke of Lorraine, inherited the region of Saint Dié and its environs. At the beginning of the second millennium, Saint Dié consisted of three groups of inhabitants: (1) Located on the side of the mountain were the churches of Saint Croix and Notre-Dame, which had fallen into disrepair. Below the churches, on the banks of the Robache River where it joins the Meurthe River, a large sandstone rock, known as Pierre Hardie, served as the place where the judges issued sentences. On the flat land adjacent to the churches, the monks lived in small houses surrounded by a granary, stables, and wine cellars. (2) Across the Robache Rapids, the farmers lived as servants of the church. (3) About a third of a mile away, on the other side of the Meurthe River, along an old Roman road, there was a suburb that included a "hospital," which attracted pilgrims as a stopover point. The attraction was based on the reputation for miraculous cures that they attributed to worshiping at Notre-Dame de Saint Dié.[6]

In the middle of the twelfth century, the monks had a wall built around an area that incorporated the two churches and the adjacent houses, and a small identifiable city was born. In 1284, under the direction of Duke Ferry II, the city was surrounded with a new wall, and impressive fortifications were constructed. Conflicts between the duke's officers and the church canons would persist over the ensuing four centuries. A walled city (see plate 1) continued to dominate the region until early in the eighteenth century.

In 1410, Saint-Dié coincidentally served as the locale where Cardinal Pierre d'Ally composed the narrative *Ymago Mundi* (Image of the World), which was published in Louvain, Belgium, in 1483. An annotated copy of that book, which Christopher Columbus carried with him during his voyages, is preserved in the Archivo General de las Indias in Seville. In 1446, a new library was built over the cloister of the cathedral, providing evidence of the chapter's dedication to learning.[7] The Battle of Nancy was fought in the region on January 5, 1477, and resulted in a glorious victory for René II, Duke of Lorraine, over Charles the Bold, Duke of Burgundy. The battle was celebrated in verse by Pierre de Blarru in "La Nancéide," the national poem of Lorraine. Neither the heroic duke, who died in 1508, nor the author of the work, who died in 1510, would witness the publication of the 5,044-line Latin poem as a beautiful book, *Liber Nanceidos*, when it came off the press of Pierre Jacobi at Saint Nicolas de Port in 1518.

In the interval between his epic military success and his death, Duke René II, who also ceremonially bore the grandiose but meaningless titles of King of Jerusalem and Sicily and Count of Provence, dedicated much of his effort to transforming Saint Dié into a cultural center that would eventually earn it the appellation Athens of Lorraine.

As part of his cultural reform, Duke René II fostered an interest in music in the region. In 1486, he had a music master, and four children were chosen for the choir. In 1498, Octavien Le Maire was designated to direct the choir, a position that he held for forty years. At the same time, two organists played the church organs. In honor of the significance of the musical efforts, the chapter had a chant book written: the elegantly illustrated *Graduel*, eight meters high and weighing forty kilograms. This tour de force continues to evoke awe today at the Room of Treasures of the Médiathèque Victor Hugo Library. The same room houses the library of Jean de Monachis, who directed the collegiate chapter during the period pertinent to the naming of America.[8]

But the most lasting secular cultural effort of the community and the one that resulted in the naming of America was particularly related to cosmography—a study of the visible universe that includes geography and astronomy. It was anticipated that the recently developed movable type printing press, which traced its origin to Johannes Gutenberg in Mainz, Germany, would play an integral role in disseminating such knowledge. Duke René II selected his secretary, Gaultier (Vautrin) Lud, to coordinate the project that led to the founding of a society of men who would meet and pool their expertise in order to produce scientific treatises that would be printed locally. The society became known as the Gymnasium Vosagense.

Gaultier Lud was born in 1448 and died in 1527. His mother was a member of a distinguished family in Saint Dié that had served the dukes of Lorraine over the years and also directed mining in the

region. Lud was made canon at Saint Dié in 1484 and adviser to Duke
René II six years later. Following in the footsteps of his brother, Jean,
Gaultier held the title secretary to the duke. After the death of his
brother, Gaultier was designated by the duke to serve as *maître général*
(general master) of the mines of Lorraine, and he also functioned as
mayor and chief constabulary for the inhabitants of the region as a
consequence of the supervisory power of the collegiate chapter of the
church of Saint Dié.[9]

The establishment of the Gymnasium Vosagense is evidence of
the effective leadership that characterized Gaultier Lud, who is
remembered for his secular contributions rather than his religious
role. He had a particular interest in the movement of the stars and
planets, and he presented his conclusions in a small treatise, *Speculi
Orbis succintiss. sed neque poenitenda neqz inelegans Declaratio et Canon*
(A most succinct description of the speculum orbis [mirror of the
world], neither difficult nor inelegant). According to the text, the
work is actually the description of an instrument, the speculum orbis,
a type of mirror. The book includes a series of fixed and movable discs
in a polar projection with one disc representing the earth on which
discs of zodiac signs and the division of the hours rotate; they are
viewed by a mobile sighting device and an alidade.[10]

This small book, consisting of a quarto of four pages, was printed
by Johann Grüninger in Strasbourg in early 1507. The only known
copy, lacking the mobile disc and map, was sold by Henry Stevens—
whose son was integral to the history of Martin Waldseemüller's world
map of 1507—to the British Library in 1864. In Gaultier's dedication
of the book to Duke René II, he states that the book consists of a
delineation of a terrestrial representation of the Ptolemaic world and
also the revolutions of the celestial spheres and seasons.[11]

Included in the book, on the third page under the heading "Speculi Fructus et Utilitates," are statements that are pertinent to the
mapping and naming of newly discovered land in the New World.
Lud refers to a figure (map) that had been hurriedly prepared of the
unknown country discovered for the king of Portugal, and writes that

a more detailed and exact representation of the coast of the country would be seen in a new edition of Ptolemy, which would soon be printed. Also, in the dedication of the work, Lud refers to Martin Waldseemüller as "the most knowledgeable man in such matters." As will be noted later (chapter 7), when the controversies that surround the Waldseemüller map are considered, Henry N. Stevens regards these declarations as evidence of the primacy of the map that he had sold to the John Carter Brown Library in 1901.[12]

Gaultier Lud's major contributions to the naming of America were the establishment of a printing press at Saint Dié and his role as recruiter of the members to the Gymnasium Vosagense and coordinator of their activities. The Saint Dié press is considered to have had as its progenitor the press of Pierre Jacobi, who, in 1501, created a printing business at Saint Nicholas de Port—a large marketplace in Lorraine. Jacobi was a priest and is credited as having been the first printer in the duchies. On his completion of the printing of *Heures de la Vierge* in 1503, he was given the title "printer of the King."[13]

In 1502, at the request of Duke René II, Jacobi undertook a consultative trip to Saint Dié to advise Gaultier Lud about developing a printing press. The press was probably not functional at Saint Dié until the latter part of 1506 or early 1507, because, if it had been available at the time Gaultier Lud's *Speculum Orbis* was printed, it is logical to conclude that the author would have used it. There is evidence that the printing press was installed in the residence of Nicolas Lud—Jean's son and Gaultier's nephew (see fig. 1). In one of the works produced by the Saint Dié press, *Novus elegansque conficiendarum epistolarum* (A new and elegant [treatise] on the composition of letters), the author, Jean Basin, states: "Generous Nicolas [here is] this little book that you have given to your workshop to print . . . illustrious Nicolas who gave the order to produce such as instructive work . . ."[14]

Additional evidence is provided by the colophon appearing at the end of *Cosmographiae Introductio*, the narrative that introduced the name "America" and the first work that came off the Saint Dié press. The initials included in the colophon—GL, NL, MI, and SD—are

Figure 1. Probable site of the original Saint Dié press.
From John Boyd Thatcher, *The Continent of America:
Its Discovery and Baptism* (New York: William Evarts Benjamin, 1896).

those of Gaultier Lud (the director), Nicolas Lud, Martin Ilacomilus (responsible author; note that the spelling is the grecized version of Waldseemüller), and Saint Dié. (See fig. 3.)

Individuals recruited to the Gymnasium Vosagense by Gaultier Lud included Hugues des Hazards, bishop of Toul; Louis de Domartin, provost of Saint Dié; Symphorien Champier, physician and author; and Johann Aluys, one of Duke René II's many secretaries. In addition to the Luds, three members played specifically critical roles in producing the narratives and graphics that contributed to the assignment of the name "America" to continental land in the New World of the Western Hemisphere.

The trio consisted of Jean Basin de Sendacour, Matthias Ringmann, and Martin Waldseemüller. The important contribution made by these particular participants is memorialized on the plaque that was unveiled at St. Dié in 1911, which states: "Here on 25 April 1507, under the reign of René II, the *Cosmographiae Introductio*, in which the new continent received the name America was printed and published by members of the Vosges Gymnasium: Gauthier Lud, Nicolas Lud, Jean Basin, Matthias Ringmann, and Martin Waldseemüller."[15]

Jean Basin de Sendacour (Joannes Basinus Sendacuius) was bishop of Wisbembach and was widely known for his expertise as a Latin scholar. His role in the development of the *Cosmographiae Introductio* was to translate the French version of the Italian original, which chronicled Vespucci's account of his four voyages, into the Latin text that appears in the printed book. This was a major factor that led to honoring Amerigo Vespucci as the discoverer of continental land in the New World. In 1507, Jean Basin became vicar at the church Notre Dame de Saint Dié and notary of the collegiate chapter. In the same year that the Saint Dié press printed the *Cosmographiae Introductio*, Basin's own little book, *Novus elegansque Conficiendarum epistolarum tractus* (A new and elegant treatise on the composition of letters), was published and dedicated to Nicolas Lud. In 1513, Jean Basin rose to secretary of the chapter and became imperial notary. He died in Saint Dié on April 21, 1523, and was buried at the collegiate church.[16]

Matthias Ringmann, who adopted as his Greek nom de plume Philesius, to which he added Vosgesigena in order to identify the general locale of his birth, was born in 1482 in Eichhoffen, a small wine-making village located in the valley of the Orbey River in the region of the Vosges Mountains. He was probably the most scholarly member of the gymnasium. His studies included theology, philosophy, Greek, poetry, mathematics, and cosmography. He was mentored by the famous philologian Jacques Wimpfeling de Schlettstadt in Heidelberg and, later, by Jacques Lefèvre d'Etaples in Paris. Ringmann contributed verses to a new edition of Albertus Eyb's *Margarita Poetica*, which was printed by Johannes Pryss in Strasbourg in 1503.[17]

Ringmann twice failed to establish a secondary school, first at Colmar and later at Strasbourg, because his teaching approach was divergent from the methods of the clerics. Consequently, Ringmann altered his career and participated in the activities of several printing establishments in Strasbourg, serving as a proofreader and editor.[18] He edited *De Ora Antarctica per regem Portugaliae pridem inuenta* (On the Antarctic region long since discovered for the King of Portugal), published in Strasbourg by Matthias Hupfuff in 1505. That work constitutes a Latin edition of *Mundus Novus*, the work attributed to Vespucci that introduced the term "New World" and includes a letter from Ringmann to his friend Jacob Braun that translates from the Latin:

> Virgil, our poet, has sung in his Æneid, that in the region beyond where the stars have their home, beyond the pathways of time and the sun, there is a land where the heaven-lifting Atlas bears upon his shoulder the celestial regions bound together by the burning stars. If one should wonder at a thing like this, he will not restrain his surprise when he reads attentively that which a great man, of brave courage, yet small experience, Americus Vespucius, has first related without exaggeration of a people living toward the south, almost under the antarctic pole. There are people in that place, he says (as you shall presently read yourself), who go about entirely naked, and who not only (as do certain people in India) offer to their king the heads of their enemies whom they have killed, but

who themselves feed eagerly on the flesh of their conquered foes. *The book itself of Americus Vespucius has by chance fallen in our way, and we have read it hastily and have compared almost the whole of it with the Ptolemy, the maps of which you know we are at this time engaged in examining with great care, and we have thus been induced to compose, upon the subject of this region of a newly discovered world, a little work not only poetic but geographical in its character.* We send to you, my friend Jacob, this work together with another book, so that you may know that you are not forgotten. Farewell. In haste, from our University, Strasburg, July 31, 1505.[19] (italics mine)

The letter provides evidence that the members of the Gymnasium Vosagense were planning a new edition of Ptolemy as early as 1505 and that they had available the narrative that described Vespucci's alleged four voyages—including an erroneously dated landing on the South American continent a year earlier than Columbus. On the recto of the second leaf of Ringmann's book, a poem relates to countries "not known, O Ptolemy, from thy maps" and the newly discovered land in the Western Hemisphere:

But afar off under the arctic pole is a certain
Land which a naked crowd of men cultivates;
This the king whom illustrious Portugal now owns
Found by sending a fleet through the shoals of the sea.
Why say more? The site and customs of the people discovered
This little book contains in a very small compass.

A variation of the poem appears in Gaultier Lud's *Speculi Orbis . . . Declaratio* that was published in Strasbourg in 1507. Intriguingly, in that work, the last line is altered to read: "The book of Americus contains in a small compass."[20]

A few historians have argued that Matthias Ringmann rather than Martin Waldseemüller, as is generally accepted, was the editor of the *Cosmographiae Introductio*. This small group of dissenting historians argues that the fact that the word "America" is deleted from the cartographic productions by Waldseemüller after 1507 might be explained if Ringmann had been the one responsible for its insertion in the *Cosmographiae Introductio* initially. But a critical issue of timing precludes crediting Ringmann with the editorship of *Cosmographiae Introductio*. It is known that at the beginning of March 1507, Ringmann was in Strasbourg, where the printing of his German translation of a work by Julius Caesar was being completed, and, therefore, it would have been impossible for him to be present at Saint Dié to edit *Cosmographiae Introductio* as it came off the press on April 25, 1507.[21] The dispute historians expressed in their assignment of credit for authorship of the text (appearing in the *Cosmographiae Introductio*) is best resolved by the assumption that the work was a product of the combined efforts of several members of the Gymnasium Vosagense, most particularly Gaultier Lud, Jean Basin, Matthias Ringmann, and Martin Waldseemüller.

The relationship between Waldseemüller and Ringmann was obviously one of mutual respect and admiration. Two of the four variant editions of the *Cosmographiae Introductio* that came off the Saint Dié press in 1507 contain a dedicatory *decastich* (ten-line poem) addressed to the emperor Maximilian by Ringmann. The same two editions also contain Waldseemüller's more personalized dedication to Maximilian.[22]

Waldseemüller's respect for and friendship with Ringmann is readily discerned in a letter that he wrote from Strasbourg to Ringmann, dated February 1508. This letter is included in Ringmann's 1508 edition of *Margarita Philosophica*, an encyclopedic compendium that first appeared in 1503, originally edited by Gregorius Reisch. The letter was composed on the occasion of Waldseemüller's contribution, titled "Architecutræ et Perspectivæ Rudimenta," to that work:

Martin Waldseemüller of Freiburg to his friend Ringmann, greeting:

In those days of the Carnival, in order to refresh myself, as is my habit, went into Germany from France, or more prop-

erly speaking, from that town in the Vosges known as St.-
Dié, where, as thou knowest, principally by my labor,
although many others, here and there, falsely claim it for
themselves, we have lately composed, drawn, and printed a
map of the entire earth, in the form of a globe as well as of
a planisphere, and which is making its way over the world
not without glory and praise. Retired for a little time in my
seclusion, while others engaged in the activities of life, I
have gathered knowledge from different authors concerning scenography, which is a branch of architecture, and
also concerning perspective, of which surely no one should
be ignorant who pretends to understand geometry. And it
is to thee, Ringmann, that at the outset I am resolved to
dedicate this, for thou thyself are most learned in the science of mathematics, thou who hadst at Paris as thy
instructor Lefèvre d'Etaples, so well skilled in all branches
of mathematics, and who now dost occupy the chair of
geography in the University of Basle, and who, moreover,
as I hear, dost also give private instruction in these subjects
to the most learned prince Christopher d' Utenheim,
bishop of Basle, great patron of students. Farewell, my
Philesius; do not cease, in caring for these other things, to
diligently follow those studies which will keep thee in the
school of philosophy.[23]

From the letter we learn that Ringmann traveled from Saint Dié
to Basel. It is also known that, in 1508, he went to Italy to borrow
some Greek manuscripts of Ptolemy; he brought them back to Saint
Dié to translate in order to provide Latin names to be included on
maps that were in production at the Gymnasium Vosagense. In 1509,
the Saint Dié press printed Ringmann's small grammar book *Grammatica figurata octo partes orationis* (Illustrated grammar in eight orations), which was illustrated with humorous drawings. Ringmann also
supplied the text for the first printed wall map of central Europe,
"Carta Itineraria Europae," which was prepared by Waldseemüller and
published in Strasbourg in 1511. That year Matthias Ringmann died

suddenly at Sélesat in Alsace, where he is buried. At the time, he was about to undertake a printing at Saint Dié of Pierre de Blarru's *Liber Nanceidos*. Although the famous edition of Ptolemy that was printed at Strasbourg was not published until 1513, it is likely that the work was essentially completed before Ringmann died. He is thought to be the author of the text of that work because his name appears twice in the publication, while Waldseemüller is not even mentioned.[24]

While it is reasonable to consider that the *Cosmographiae Introductio* had input from several members of the Gymnasium Vosagense, and that Matthias Ringmann contributed significantly to the incorporation of the Vespuccian material, the one participant who stands above all others in the production of the work is generally accepted to be Martin Waldseemüller. Not only did he most likely coordinate the production of the text, but he is indisputably the man who made the two maps that display the name "America."

Martin Waldseemüller (see frontispiece) was born sometime between 1470 and 1475, either in Wolfenweiler near Freiburg im Breisgau, Germany, or in Radolfzell on the shore of Lake Constance. His father, Conrad, was a successful butcher who moved to Freiburg shortly after Waldseemüller's birth and served as bailiff and treasurer of that town. The first authenticated information about the man who made the world map of 1507 consists of the matriculation register of the University of Freiburg, where Waldseemüller's name appears on December 7, 1490, as "Martinus Walzenmüller de Friburgo Constantiensis Diocesis." The date of matriculation does not preclude that he was born even later than 1475 because, at the time, many students began their university studies at an early age. Johann Schött, who went on to print the 1513 Ptolemy and may have served as a printer at Saint Dié during the activities of the Gymnasium Vosagense, matriculated at Freiburg as a thirteen-and-a-half-year-old on the very same date as Waldseemüller.[25]

There is no documentation of the specific subjects that Waldseemüller studied, but the classic educational curriculum at the time was the "trivium" of grammar, rhetoric, and logic, followed by the "quadrivium" of arithmetic, music, geometry, and astronomy. At some point, Waldseemüller interrupted his studies to work in his uncle's printing establishment in Basel. He returned to the university, studied theology, and became an ordained priest in the Diocese of Constance.

Waldseemüller's subsequent presence is not chronicled until he arrived at Saint Dié in 1506 as a member of the Gymnasium Vosagense. The assumption is that he was recruited by Gaultier Lud for his expertise in geography and cartography and for his experience with printing. Perhaps Waldseemüller had gained cartographic experience and participated in the process of printing maps in the shop of Johann Grüninger in Strasbourg.

In the environment of the gymnasium, Waldseemüller followed the practice of other humanists, particularly his friend Matthias Ringmann—Philesius—by adopting the scholarly appellation Hylacomylus, or Ilacomilus, the Greek literal translation of his German name, which means "miller of the lake in the forest."[26] It was Alexander von Humboldt, the famous German naturalist and explorer, and the author of the first scientific work about the South American continent, who established the identity of Hylacomylus as Martin Waldseemüller and postulated his authorship of *Cosmographiae Introductio*.

Washington Irving, the first American author to gain international acclaim, wrote in a "Note" to his 1848 revised edition of *The Life and Voyages of Christopher Columbus*:

> Humboldt in his EXAMEN CRITIQUE, published in Paris in 1837, says: "I have been so happy as to discover, very recently, the name and the literary relations of the mysterious personage who (in 1507) was the first to pose the name of America to designate the new continent, and who concealed himself under the Grecianized name of "Hylacomylas." He then, by a long and ingenious investigation, shows that the real name of this personage was Martin Waldseemüller, of Fribourg, an eminent cosmographer, patronized by René, Duke of Lorraine, who, no doubt, put in his hand the letter received by him from Amerigo Vespucci. The geographical works of

Waldseemüller, under the name of Hylacomylas, had a wide circu-
lation, went through repeated editions, and propagated the use of
the name America throughout the land. There is no reason to suppose that this application of the name was in anyways suggested by
Amerigo Vespucci. It appears to have been entirely gratuitous on
the part of Waldseemüller.[27]

Waldseemüller's involvement and probable leadership role in the
production of the *Cosmographiae Introductio* are supported in his
printed statement (translated):

> It thus happened that in collecting for my own work; aided by some
> others, the books of Ptolemy and collating them with the Greek
> text, and in proposing to add thereto an inquiry into the four voy-
> ages of Americus Vespucci, I have prepared for the common use of
> students, and as a sort of preparatory Introduction, a figure of the
> entire earth, as well in the form of a globe as a representation on a
> flat surface; and I have dedicated it to Your Most Sacred Majesty,
> who holdeth in his hand the empire of the world. Hoping that my
> wish be realized, and that under the shelter of thy shield (as of that
> of Achilles) I would be safe from the designs of the envious, should
> I be able to satisfy, at least in part, the judgment of Your Majesty, so
> discriminating in these matters. Hail illustrious Emperor. From the
> city of St.-Dié in the year after the birth of our Lord, 1507.[28]

Although the sequence of events is unknown, it is likely that the
initial plan of the group that had been assembled at Saint Dié was to
produce and print a new edition of the Ptolemy atlas, the last two
having been published in Ulm, Germany, in 1486 and in Rome in
1490. The direction of the members of the Gymnasium Vosagense
changed when they learned of the purported accomplishments of
Amerigo Vespucci and, more significantly, when they received a
French translation of *Lettera di Amerigo Vespucci delle isole nuovamente
trovate in quattro suoi viaggi* (Letter of Amerigo Vespucci concerning
the isles newly discovered on his four voyages). This is a twenty-four-

page printed document in which Vespucci described four voyages that he allegedly conducted to the New World—a term that had been introduced in *Mundus Novus*, a 1504 narrative also ascribed to Vespucci. The result was the production of the *Cosmographiae Introductio* and the two accompanying maps. The text was first published at Saint Dié on April 25, 1507, and it was followed by three other editions that year.[29]

Unquestionably, it was Waldseemüller who made both the large "plane" 1507 world map and the 1507 gore map (a map to be fitted on a sphere to make a globe), both of which are defined on the title page of the *Cosmographiae Introductio* as being integral components of that work. It is thought that the maps were actually printed in Strasbourg, most likely by Johann Grüninger, where the remainder of Waldseemüller's cartographic works were produced.[30]

In 1508, Waldseemüller contributed an illustrated chapter on architecture and perspective to the fourth edition of *Margarita Philosophica*. In 1511, he published the first printed wall map of central Europe, "Carta Itineraria Europae," to which Matthias Ringmann contributed a descriptive text titled "Instructio manuductionem prætans in Cartam Itinerium Martini Hilacomili." This large map, measuring 141 × 107 centimeters (56 × 42 inches), was published by Johann Grüninger in Strasbourg and depicts the most important trade routes of the period. The streets are shown as dotted lines; hills and mountains as molehills; and the woods are marked with trees. It was dedicated to Duke Antonio, the son and successor of Duke René II (who died in 1508), with the hope of interesting him in future cartographic projects. The only known example is preserved in the Ferdinandeum State Museum of the Tyrol (Tiroler Landesmuseum) in Innsbruck, Austria.[31]

The 1513 edition of Ptolemy's *Geographia* that had been conceived of and initiated at Saint Dié in 1505 or 1506 was eventually published in Strasbourg by Jacobus Eszler and Georgius Übelin, with Johann Schött as its printer. The work includes two new maps of the New World, "Orbis Typus Universalis Iuxta Hydrographorum Traditionem" and "Tabula Terre Nove," both ascribed to Waldseemüller. He also produced the first printed detailed maps of Switzerland, Lorraine, and the Upper Rhine, among a total of twenty modern maps created for that edition. In 1514, Waldseemüller was appointed cleric of the Diocese of Constance for the canonry at Saint Dié, but his clerical commitments did not prevent him from continuing his cartographic productivity. The 1515 edition of Gregorius Reisch's *Margarita Philosophica* contains Waldseemüller's own larger and more detailed version of his "Typus Orbis" map that he had produced for the 1513 Ptolemy *Geographia*.[32]

Also, in the 1515 edition of *Margarita Philosophica*, Waldseemüller described and depicted his polimetrum. This instrument was a forerunner of the theodolite (a portable instrument for measuring horizontal and vertical angles). It consisted of "a horizontal plate with a vertically suspended semicircle and equipped with index arms, leads, surveying quadrants, and graduated arcs." The name *polimetrum* does not appear in any subsequent scientific literature.[33]

In 1516, Waldseemüller's monumental "Carta Marina Navigatoria Portugallen Navigationes Atque Tocius Cogniti Orbis Terre Marisque" was published in Strasbourg. Waldseemüller was later approached by Johann Grüninger to modify the "Carta Marina" by adding German inscriptions and a fully illustrated German text, but death prevented Waldseemüller from completing the project that was taken over by Laurentius Fries.

Bedini states that Waldseemüller died in his canon house at Saint Dié on March 16, 1519. Ronsin gives the year as 1520. The only established point of reference indicates that Waldseemüller's death occurred before March 12, 1522, because, on that date, a document states that the vacant canonry at Saint Dié was filled. Also, the 1522 edition of Ptolemy honors him with the statement "Lest we seem to claim the merits of others, we declare that these maps were originally constructed by Martin Waldseemüller piously deceased . . ."[34]

The press at Saint Dié was short lived and minimally productive. On April 25, 1507 (VII Calends May), two variations of an edition of

Cosmographiae Introductio were published as the press's initial production, and, on August 29, 1507 (IIII Calends September), two additional editions of the book came off the press. The other publications of the press at Saint Dié were probably completed subsequently in the following order. The first was Jean Basin de Sendacour's *Novus elegansque Conficiendarum epistolarum tractus*. No copy of this book has been found, but there is evidence that the historian Johann Daniel Schöpflin owned one. The next publication was Matthias Ringmann's *Grammatica figurata octo partes orationis*, which bears a date of 1509. That same year, *Chartiludium logice seu logica poetica vel memoritiva* (The logic of card games, poetic or mnemonic logic) by Thomas Murner was reprinted from the original that had been published at Kraków in 1507. Once again, it is known that a copy was in Schöpflin's library in Strasbourg, and it was lost when that library was destroyed by fire. Gaultier Lud dedicated the book to Bishop Hugues des Hazards.[35]

A quarto of six leaves, titled *Renati secundi Sicilæ regis Lotharingiæ ducis vita per Johannem Aluysium Crassum Calabrum edita* (The life of René the Second, King of Sicily and Duke of Lothargia edited by Joannes Alusius Crassus of Calabria), is dated 1510 and focuses on the life of Duke René II two years after he died. A copy is in the library at Schettstadt. That small work and a quarto of twelve leaves, *Defensio Christianorum de Cruce id est: Lutheranorum* . . . (A defense of Christian [authors] concerning the cross: or of the Lutherans), both published in 1520, complete the entire corpus of production of the original Saint Dié press. It is believed that the press was transferred to Johann Schött in Strasbourg shortly before Gaultier Lud died in 1527. More than a century would pass before another press was established at Saint Dié. In 1625, Jacques Marlier used the new press to publish a small quarto in French, *Recherches de Saintes Antiquitéz de la Vosges* (Investigations concerning the Ancient Saints of the Vosges).[36]

Thus, in a remote locale, in what at the time was known as the Duchy of Lorraine, with the sponsorship of an enlightened member of the aristocracy, under the direction of a recruiter of talent, a group of mainly ordained priests used a relatively recent invention, the printing press with movable type, to spread knowledge. All but one of the textual productions of the original Saint Dié press have faded into obscurity. But, that one item—a small book—set in print the narrative that would present the basis for the naming of America. The title page of that work announced the production of one map, gores for a globe, and a flat map, now know as the monumental Waldseemüller world map of 1507.

CHAPTER 2
BACKGROUND AND BASES

Not the possession of the truth but the effort in struggling to attain it brings joy to the researcher.

Gotthold Lessing

The products of the Gymnasium Vosagense integral to the placement of "America" on land in the Western Hemisphere consist of two main elements: a declarative text—*Cosmographiae Introductio*, and the accompanying two maps—a small set of gores for a globe and a large plane map with a size and elegance that is considered meritorious of its appellation as America's Birth Certificate or Baptismal Document. The textual material is known to have initially come off a small press at Saint Dié on April 25, 1507. The conception and execution of the two maps undoubtedly took place at that same locale, but it is probable that their printing was carried out by a more sophisticated press at Strasbourg. In any event, the material that led to the naming of America did not evolve in a historic vacuum. Rather, it was based on the writer's (writers') and cartographer's knowledge and an appreciation of preceding events, discoveries, narratives, and maps.

History is written under diverse and complex circumstances. Occasionally, it is the report of an active participant or a witness. Even in this situation, interpretation is a crucial factor, as portrayed in the Japanese film *Rashomon*, in which the two participants and one witness present three distinctly disparate scenarios. More often, history is prepared by a scholar working at a distance in time and place, and, therefore, is wholly dependent on previous reports. This, of course, compounds the interpretative process. It depends on the historian's sifting through alleged facts, accepting some, rejecting others, and modifying several. History is thus defined, in part, by the historian. Similarly, maps are occasionally produced directly by the artistry of those who personally viewed and interpreted the geography. But, more frequently, a particular map is made by perpetuating or interpreting previous narratives and graphic presentations. Thus, representative geography is defined, in part, by the cartographer.

Therefore, before considering the documents that played dominant roles in the naming of America, the antecedent historic events and discoveries that might have contributed to the statements and pictures made in those documents should be brought into focus. In addition, the previous productions, both as manuscripts and printed matter and appearing as text or graphic representations, which were or could have been available for assessment by members of the Gymnasium Vosagense, should be defined.

A chronological consideration of the events that might have been influential to the thinking concerning the New World that took place at the Gymnasium Vosagense in Saint Dié early in the sixteenth century spans a brief period of, at most, fifteen years, from 1492 to 1507. During that time, a few voyages of exploration to the Western Hemisphere provided information that could have reached the members of the gymnasium who were involved in the production of the *Cosmographiae Introductio* and the two accompanying maps.

The initial event occurred when Christopher Columbus, a Genoese sea captain, sailing under the flag of Spain and the patronage of King Ferdinand and Queen Isabella, undertook what he termed

"Empresa de las Indias" (Enterprise of the Indies). At the time, the European term "The Indies" applied to the vast region including China, Japan, the Ryukyu Islands, the Spice Islands, Indonesia, Thailand, and everything between them and India. Columbus proposed to sail in a westerly direction, which he believed would provide a short and direct route to these prized lands by obviating the need to sail around the tip of Africa.[1]

Columbus based his vision, in large part, on a letter by the Florentine physician/cosmographer Paolo dal Pozzo Toscanelli. A copy, dated June 25, 1474, of the original letter written by Toscanelli to Fernão Martins of the cathedral of Lisbon was sent to Columbus, by its author, accompanied by a chart that Columbus carried with him on his voyages. The appeal of Toscanelli's thesis, as depicted on the map, was based on a gross miscalculation, which estimated the distance between the Canary Islands and Japan at 3,000 nautical miles instead of what it actually is, about 10,600 miles (when measured at 28° north latitude). Columbus, whose copy of the map remains in Biblioteca Colombina, Seville, actually estimated that distance to be 2,400 miles. This miscalculation erroneously placed Japan at a distance from Spain equivalent to the actual distance between Spain and the West Indies.[2]

On August 2, 1492, Columbus's fleet of three ships, the Niña, the Pinta, and the Santa Maria, set sail from Palos, Spain. On October 12, the crew first spotted land in the Western Hemisphere. Leaving his flagship, the Santa Maria, and accompanied by Captain Vicente Yáñez Pinzón of the Niña and Vicente's brother, Captain Martin Alonso Pinzón of the Pinta, Columbus set foot on an island that the natives called Guanahani and that Columbus named "San Salvador" for the Holy Savior. During Columbus's first voyage, sightings or landings were made at islands that were named by the explorers at the time; "Santa Maria de la Concepción" (Rum Cay), "Fernandina" (Long Island), "Isabela" (Crooked Island), Cuba (called Colba by the natives and named "Juana" for the Spanish Infanta by Columbus), and "La Isla Española" (Haiti/Dominican Republic). On Christmas

Day, the *Santa Maria* was wrecked on a coral reef off the north shore of Hispaniola. A settlement was established on the island (currently located in Haiti) near the wreck and named "Villa de la Navidad." Twenty-one crew members remained at that settlement, while the *Niña* (with Columbus aboard) and the *Pinta* departed for home on January 4, 1493.

On March 4, bad weather and heavy seas forced Columbus, aboard the *Niña*, to land at Lisbon. In order to dramatize the event, Columbus brought along captive Indians when he called on the king of Portugal, Dom João II. Columbus then proceeded to Palos in Spain, finally arriving on March 15, 224 days after his voyage began. Columbus's first voyage was reported in his "Letter to Santángel," treasurer of Aragon. The original letter, the only copy of which is a treasure of the New York Public Library, is a four-page document that was printed in Barcelona in April 1493. The first illustrated copy was printed later in Basel in 1493, containing illustrations of a vessel and an allegoric landing of Columbus.

Columbus was greeted by Ferdinand and Isabella in April at Cordova, where the sovereigns bestowed upon him the title "Admiral of the Ocean Sea, Viceroy and Governor of the Islands he hath discovered in the Indies." On September 25, 1493, in command of a large fleet, indicative of the glory that he enjoyed, Columbus departed Cadiz on his second voyage. On November 3, landfall was made in the Lesser Antilles on an island that he named "Dominica." The fleet continued on to an island that Columbus named "Santa Maria de Guadalupe" (Guadeloupe), where the ships remained for several days. Following a northwesterly course, the ships sailed on the leeward side of the islands, which were named "Santa Maria de Monserrate" (Montserrat), "Santa Maria la Antigua" (Antigua), "San Martin" (Nevis), "San Jorge" (St. Kitts), and "Santa Anastasia" (St. Eustatius).

The fleet then turned west and reached the north shore of St. Croix, an island that the natives called "Ayay." Continuing in a westerly direction, the ships sailed south of the Virgin Islands and the south shore of Puerto Rico, which was called "Boriquén" by the natives and "San Juan Bautista" by Columbus. According to Samuel Eliot Morison, on about November 20, a landing took place at Puerto Rico on the shores of Añasco Bay rather than Boqueron Bay, the site designated by previous historians. Several days later, the fleet reached Hispaniola, where Columbus learned that the settlement, which had been left at Navidad during the first voyage, had been completely wiped out by disease and the natives. In January 1494, to the east of Caracol Bay, Columbus established the first European colony in the Western Hemisphere that survived. He named it "Isabela."[3]

Shortly after that colony was established, Alonso de Hojeda, who would later play a significant role in several subsequent voyages to the Western Hemisphere—including the first voyage of Amerigo Vespucci that was critical in the naming of America—led an inland exploration to contain the natives. Columbus elected to retain only three large vessels and two smaller caravels at Hispaniola; the other twelve vessels were sent back to Spain where they arrived on May 7, 1494. While the main portion of the expedition returned to Spain, Columbus sailed with three caravels from Isabela on April 24, 1494, and landed on the southeastern tip of Cuba, which he thought was the Chinese province of Mangi. The small fleet continued along the southern shore in a westerly direction and entered a harbor that they named "Puerto Grande" (currently Guantánamo Bay). They proceeded west to "Cabo de Cruz" and then turned south, entering a harbor on the north shore of Jamaica on May 5, 1494.[4]

The fleet sailed west along the north shore of Jamaica for several days and then turned north to continue an exploration of Cuba in search of evidence of Chinese culture. In mid-June, the ships reversed course and, in July, sailed into Montego Bay, Jamaica. After sailing around the western tip of Jamaica, the fleet followed the south shore of the island and continued eastward until September 29, 1494, when they reached Hispaniola at the point from which the voyage had originated. On his arrival, Columbus was surprised to be met by his brother, Bartholomew, who had brought provisions from Spain. More provisions arrived at the end of the year.

Discontent among the natives was pervasive. Led by Christopher Columbus, Bartholomew, and Columbus's other brother, Diego, the crew members thwarted an attempt by the natives to unite against the Spaniards. During the process, Columbus captured about fifteen hundred natives and shipped many of them to Spain. By 1496, the native population was entirely subdued. The atrocities perpetrated under Columbus's leadership were reported to Ferdinand and Isabella, who, in response, sent an envoy to investigate the situation. In the midst of the investigation, Columbus departed Isabela on March 10, 1496, leaving his brother Bartholomew in charge. After a short stay in Guadeloupe, the second voyage of Columbus ended at Cadiz on June 11, 1496.[5]

Columbus planned his third voyage, in part, to search for the mainland south of the islands. He had heard from the king of Portugal, Dom João II, that a mainland existed in this region. Because the Portuguese king believed such land existed, he insisted that the papal line of demarcation, established by the Treaty of Tordesillas in 1494 to divide the land in the New World between Spain and Portugal, should be moved farther west so that more land would be designated to Portugal. Columbus himself had received hints of such a mainland from the natives that he had encountered previously. Therefore, it was planned that three vessels were to sail directly to Hispaniola with settlers and provisions, while Columbus was to command three additional caravels on a voyage specifically for discovering new lands.[6]

But before Columbus embarked on his third voyage, John Cabot, sailing under the flag of England, made landfall in the northern part of the Western Hemisphere. Cabot's voyages are shrouded in mystery. All evidence related to Cabot's voyage of discovery is extrapolated from the correspondence of others. There is no authentic description of Cabot, no precise knowledge of his background or prior activities, and no journal of his voyage.

John Cabot was an Italian and, probably, a Genoese with the name of Giovanni Caboto, or, perhaps, Gaboto, Cabuto, or another variant. There is a record that he became a naturalized citizen of Venice in 1476 after residing there for fifteen years. From a letter written on December 18, 1497, by Raimondo Soncino, the Milanese envoy to London, we learn that he was "a very good mariner. . . . Of a fine mind, greatly skilled in navigation."[7]

The Soncino letter, coupled with a letter from the English merchant John Day written in Andalusia to the "Lord Grand Admiral" of Spain (presumably Columbus), in addition to questionable assertions by Cabot's son, Sebastian, provide the only contemporary references to the first English voyage of exploration. Day mentions an earlier abortive attempt by John Cabot to sail west from England and other Atlantic expeditions conducted by Bristol seamen. The letter, the original manuscript of which is in the Archivo General de Simancas, Estado de Castilla, Leg.fol.6 (translation L. A. Vigneras), states:

From the said copy your Lordship will learn what you wish to know, for in it are named the capes of the mainland and the islands, and thus you will see where land was first sighted, since most of the land was discovered after turning back. Thus your lordship will know that the cape nearest to Ireland is 1,800 miles west of Dorsey Head which is in Ireland, and the southernmost part of the Island of the Seven Cities is west of Bordeaux River, and your Lordship will know that he [Cabot] landed at only one spot on the mainland, near the place where land was first sighted, and they disembarked there with a crucifix and raised banners with the arms of the Holy Father and those of the King of England, my master; and they found tall trees of the kind masts are made, and other smaller trees, and the country is very rich in grass. In that particular spot, as I told your Lordship, they found a trail that went inland, they saw a site where a fire had been made, they saw manure of animals which they thought to be farm animals, and they saw a stick half a yard long pierced at both ends, carved and painted with brazil, and by such signs they believe the land to be inhabited. Since he was with just a few people, he did not dare advance inland beyond the shooting distance of a cross-bow, and after taking in fresh water, he returned to his ship. All along the coast they found many fish like those which in Iceland are dried in the open and sold in England and other countries, and these fish are called in England "stockfish"; and thus following the shore they found two forms running on land one after the other, but they could not tell if they were human beings or animals; and it seemed to them that there were fields where they thought might also be villages, and they saw a forest whose foliage looked beautiful. They left England toward the end of May, and must have been on the way 35 days before sighting land; the wind was east-north-east and the sea calm going and coming back, except for one day when he ran into a storm two or three days before finding land; and going so far out, his compass needle failed to point north and marked two rhumbs [lines crossing successive meridians at the same angle] below. They spent about one month discovering the coast and from the above mentioned cape of the mainland which is nearest to Ireland, they returned to Europe in fifteen days. They had the wind behind them, and he reached Brittany because the sailors confused him, saying that he was heading too far north. From there he came to Bristol, and went to see the King to report to him all the above mentioned; and the King granted him an annual pension of twenty pounds sterling to sustain himself until the time comes when more will be known of this business, since with God's help it is hoped to push through plans for exploring the said land more thoroughly next year with ten or twelve vessels—because in his voyage he had only one ship of fifty "toneles" and twenty men and food for seven or eight months—and they want to carry out this new project. It is considered certain that the cape of the said land was found and discovered in the past by men of Bristol who found "Brasil" as your Lordship well knows. It was called the island of Brasil, and it is assumed and believed to be the mainland that the men from Bristol found.

Since your Lordship wants information relating to the first voyage, here is what happened: he went with one ship, his crew confused him, he was short of supplies and ran into bad weather, and he decided to turn back.[8]

It can be inferred that John Cabot received a patent for discovery from King Henry VII in 1496, at which time he sailed west from Bristol but turned back shortly after departure because of storms and disagreements with his crew. As his 1497 first voyage of discovery is reconstructed, Cabot left Bristol in command of the *Matthew* at the end of May and departed Dorsey Head, Ireland, westward on a passage of thirty-two or thirty-three days. Landfall was made on June 24, 1497, according to an inscription on Sebastian Cabot's map of 1544.[9]

According to Samuel Eliot Morison's deductive reasoning, Cabot sighted an island that he named "St. John's" after the saint whose canonization was celebrated on June 24; the French later gave it its current name, Belle Isle. Cabot is thought to have made his landing south of Cape Dégrat on Quirpon Island, which is separated from Newfoundland by a narrow strait. As is stated in the letters of John Day (see above) and Raimondo Soncino, "After having wandered for some time, he, at length arrived at the mainland, where he [Cabot] hoisted the royal standard and took possession for the king here." Cabot formally staked a claim for England.[10]

The ship continued south along the east coast of Newfoundland, sailing within the Grand Banks where, according to Soncino, "[t]hey assert that the sea is swarming with fish, which can be taken not only with the net, but in baskets let down with a stone, so that it sinks in the water. I have heard this Messer Zoane [John Cabot] state so much." After returning to the point of original landfall on July 20, the *Matthew* began her voyage home, arriving in Bristol on August 6, 1497. A few scholars have conjectured that Cabot might have sailed as far north as Labrador, and that the trees described in the Day letter indicate that he actually also reached the coast of Maine. In any event, Cabot thought that he had reached Asia, "the country of the great Khan" or the "Island of the Seven Cities."[11]

Upon his arrival at Bristol, Cabot immediately left for London to report his finding to the king and, according to Soncino, presented a map that has disappeared. That map, which certainly depicted Newfoundland as part of the Eurasian continent, found expression on the 1500 map by Juan de la Cosa. All that is known of Cabot's subsequent and final voyage is that five ships departed Bristol in May 1498, after which only one ship returned. The remainder of the fleet, including Cabot himself, was lost.[12]

John Cabot's son, Sebastian, created a world map in 1544, which exists as a lone copy in the Bibliothèque Nationale in Paris. The statements made in the inscriptions on the copper engraving were not written by Sebastian, who is regarded by two distinguished historians, Henry Harrisse and Samuel Eliot Morison, as a totally unreliable source; and the accuracy of both the inscriptions and Sebastian Cabot's authorship of the map have been questioned. The map gives the date and hour of landfall and describes the country discovered by "Juan Cabot, a Venetian, and by Sebastian, his son," and places the "land first seen" at Cape Breton, on Cabot Strait. Sebastian was fifteen at the time that he supposedly accompanied his father on that voyage.[13]

Subsequent to Cabot's voyages, Christopher Columbus began his third voyage on May 30, 1498, at the mouth of the Guadalquivir River near Seville. The three caravels led by Columbus on that voyage of discovery sailed to the Cape Verde Islands, from which they proceeded westward and made landfall at Trinidad on July 31, 1498. Columbus named the land "la ysla de Trinidad" for the three mountains or large rocks that he viewed from his ship. After crossing the Gulf of Paria, on August 1, Columbus sighted part of modern-day Venezuela, that is, continental land for the first time. Nevertheless, because he thought it was an island, he named it "Ysla Sancta." The explorers first set foot on continental land when they went ashore on the south shore of the Paria Peninsula on August 5, 1498. This was the first time any European touched continental land in the Western Hemisphere since the Norse voyages. John Cabot can be credited only with an island landfall near the North American continent at Quirpon Island near Newfoundland in 1497. While coasting close to the shore, Columbus's crew sighted the mouths of four rivers, including what later received the names Rio Grande and Orinoco. According to the captain of one of the caravels in Columbus's fleet, Columbus himself went ashore and took possession of the land in the names of the sovereigns of Spain.[14]

Columbus did not recognize that he had set foot on continental soil; he still believed that he had encountered yet another island. The small fleet sailed on in a westerly direction north of Venezuela to "Margarita Island," which Columbus named for the Infanta Margarita of Austria. While on that course, Columbus sighted the island of Grenada, which he named "Asunción," because it came into view on the Day of the Feast of Assumption. Finally, on August 15, Columbus concluded that he had seen the mainland and recorded in his journal: "I believe that this is a very great continent, until today unknown. . . . And if this be a continent, it is a marvelous thing, and will be so among all the wise, since so great a river flows that it makes a fresh-water sea of 48 leagues." Columbus referred to these lands that he discovered as "otro mundo," or "Other World." Columbus's Other World, however, did not mean a New World as we now think of it as indicating the Western Hemisphere, nor, in fact, did Vespucci's *Mundus Novus*. Rather, these terms referred to a world not mentioned previously in Ptolemy's *Geographia*. To his dying day, Columbus believed that the "continent" he encountered was part of China and the Orient.[15]

Columbus arrived at Santa Domingo, the new capital of Hispaniola, located on the south shore, on the last day of August 1498. Santo Domingo became the first significant permanent settlement of Europeans in the Western Hemisphere. During the ensuing period, the administrative performance of the three Columbus brothers, Christopher, Bartholomew, and Diego, alienated most of the Spaniards on the island. As a consequence, the Spanish sovereigns dispatched a commission to the island. After the situation was reviewed, the brothers were sent home for trial. On New Year's Day 1501, Columbus was replaced as governor of Hispaniola. Nevertheless, Ferdinand and Isabella did eventually agree to allow Christopher Columbus to lead a fourth voyage to the Indies.

One other voyage was reported to have taken place before the fifteenth century came to an end. It was a voyage that would inappropriately play a significant role in the naming of America. It was referred to in the published *Lettera* attributed to Vespucci as the first of his four voyages, which erroneously reported a landing on the Paria Peninsula of Venezuela in 1497 rather the actual date of 1499.

In May 1499, Alonso de Hojeda—who captained a caravel on Columbus's second voyage and led troops in battles against insurgent natives on Hispaniola—departed Puerto de Santa Maria near Cadiz in command of four ships. The expedition was licensed by Bishop Don Juan de Fonseca and financed by several merchants in Seville. Sailing with Hojeda were Juan de la Cosa—who would later draw the first map to depict the discoveries of Columbus and Cabot—and Amerigo Vespucci. Vespucci's role on the expedition was relatively minor and ill defined, and not much is known of his activities prior to the voyage except that he was employed as a Florentine merchant-banker while a resident in Seville.[16]

After stopping in Morocco and the Canaries, the transatlantic crossing brought the small fleet to the Western Hemisphere. The ships sailed to the Paria Peninsula, where Columbus had landed in 1498, and continued westward to Margarita Island where some crew members went ashore. The ships then proceeded to an island that they named "Gijantes" (now Curaçao) after the tall natives the crew witnessed at that location. They noted a village on the mainland in the region of the Gulf of Maracaibo with houses built over the water on stilts and, because it reminded them of Venice, they named it "Little Venice," or Venezuela. Hojeda probably set foot in South America on the shore opposite Curaçao, before continuing westward around the Guajira Peninsula to Cabo de la Vela, and then turning north to the southwestern shore of Hispaniola. The expedition returned to Spain in the late spring or early summer of 1500.[17]

Shortly after Hojeda sailed from Palos, in 1499, Peralonso Niño and Cristóbal Guerra also left that port and arrived in the New World at about the same time as Hojeda. After trading on the Pearl Coast for three months, they sailed to the Guajira Peninsula of Venezuela and then returned to Spain.

In the early part of 1500, Vicente Yáñez Pinzón, who commanded the *Niña* during Columbus's first voyage, discovered northern Brazil and the mouth of the Amazon River. Pinzón's voyage of exploration began on November 18, 1499, when he set forth from Palos with four caravels. In the latter part of January 1500, the fleet made landfall at 8° south latitude, where he named a cape (near modern-day Recife in Brazil) "Santa Maria de la Consolación." The Portuguese previously had named the same area "San Agostinho." Pinzón then sailed west and northwest to the Gulf of Paria. During this segment of the voyage, he entered the Amazon River, which he named "Marañon." After crossing the Gulf of Paria, he sighted Tobago, which he named "Isla de Mayo," then proceeded to the Bahamas, which he had visited in 1492, refitted at Hispaniola, and returned to Palos on September 30, 1500.

As the sixteenth century began, there was a flurry of expeditions to the New World. On March 8, 1500, Pedro Álvares Cabral led a fleet of thirteen or fourteen Portuguese vessels from the mouth of the Tagus River at Lisbon. They sailed past the Cape Verde Islands in a southwesterly direction, making landfall in South America on April 22, 1500, at the mouth of the "Rio Cahy," about 17° south latitude, where they sighted and named "Monte Pascoal." Cabral thought he came upon an island and named it "Ilha de Vera Cruz." Cabral dispatched the store ship home to inform King Dom Manuel of the discovery, which is recorded on the 1500 *mappemonde* (world map) of Juan de la Cosa and the Cantino map of 1502. The fleet sailed north along the east coast of Brazil for about forty miles before it changed course and proceeded to India, eventually returning to Lisbon in the summer of 1501. On July 5, 1500, Rodrigo de Bastidas left from Seville with two caravels and reached Cabo de la Vela on the north coast of South America, and then sailed along almost the entire coastline of Colombia, including Santa Marta, Barranquilla, and Cartagena.[18]

Alonso de Hojeda, with a license from Bishop Fonseca, led another expedition, consisting of four vessels, to the so-called Pearl Coast of northern Venezuela in 1502; he established a trading post on the Guajira Peninsula. Hojeda died in Santo Domingo, Hispaniola, in 1515. His explorations were recorded on the famous Juan de la Cosa mappemonde of 1500, drawn by Alonso de Hojeda's shipmate. Juan de la Cosa himself made five or six voyages to the Western Hemisphere before he met his death there as a result of a poisoned arrow on February 28, 1510.[19]

Meanwhile, to the north, sometime in the late spring or early summer of 1500, Gaspar Corte-Real left Lisbon and, after the Atlantic crossing, made landfall at about 50° north latitude on "land that was very cool and with big trees," which he named "Terra Verde." It was probably Newfoundland. Gaspar set out on a second voyage to the same region in mid-May 1501, but was lost at sea. His brother Miguel, who embarked in May 1502 on a voyage to Newfoundland in an attempt to find Gaspar, suffered the same fate.[20]

The alleged four voyages of Vespucci that are detailed in the printed *Lettera*, a Latin translation of which was incorporated in the *Cosmographiae Introductio*, will be considered extensively in the next chapter. Included in *Lettera*, in addition to the first voyage aboard Hojeda's ship that was misdated 1497 and took place in 1499, two other trips, which might have taken place, are described. These are reported to have extended along the coast of South America as far south as Cape St. Augustine at 8° south latitude.

To complete an examination of the explorations in the Western Hemisphere that could have influenced mapmakers prior to 1507 and, more particularly, members of the Gymnasium Vosagense in their preparation of the text and accompanying two maps that named America, we end with the same explorer with whom we began in 1492.

With authorization and funding from sovereigns Ferdinand and Isabella, Christopher Columbus, commanding a fleet of four caravels, set sail from Cadiz on May 9, 1502. After stopping and taking on provisions at Grand Canary, the ships proceeded westward and southward, reaching Martinique on June 15. Lasting only twenty-one days, this was Columbus's fastest transatlantic crossing. Several days later, the vessels called at Dominica, and then continued to Santo Domingo, where they were not allowed entrance into the harbor. This was in accordance with the sovereigns' forbiddance of Columbus visiting until the fleet was homeward bound.[21]

About thirty miles north of the coast of Honduras, the ships reached the island of Bonacca (called "Maia" by the Jicaque natives) on July 30. In early August, Christopher Columbus once again set foot on continental land in the Western Hemisphere, this time at Cape Honduras, which he named "Punta Caxinas" for the Arawak name of a local tree. On August 14, the ships anchored off the mouth of a river named "Rio de la Posesiòn" because it was there that Columbus took formal possession of the land for Spain; it is now called Rio Romano. Continuing east along the north coast of Honduras, the ships reached the most eastern point, Cabo Gracias á Dios, and then followed the east coast of Nicaragua in its entirety and, subsequently, the east coast of Costa Rica to Puerto Limón. The fleet completed the coastal journey by sailing east along the north coast of Panama, at first entering the Chiriqui Lagoon. Up to this point, Columbus was specifically conducting a search for a strait that would provide access to the west.[22]

Columbus and his son spent more than three months, including Christmas and New Year's Day, in what is now the harbor at the Atlantic entrance of the Panama Canal. After coasting back and forth along the north shore of Panama, they sailed outside of the Archipelago de los Mulatos to its most eastern point and then turned north for Hispaniola with a reduced fleet of three vessels on Easter night, April 15, 1503. After brief stops at Cuba then Jamaica, they finally arrived at Hispaniola. Columbus, his son, his brother, and twenty-three crew members left that island for Spain on September 12, 1504, arriving on November 7, more than two years after the beginning of Columbus's final voyage of exploration.[23]

Individually or cumulatively, these voyages might have influenced Waldseemüller directly. As an alternative, he might have been influenced by the previous maps to which he was exposed before 1507. These voyages and their resulting maps had defined most of the islands in the Caribbean Sea, a portion of Newfoundland, the northern coast of Venezuela, the coasts of Honduras, Nicaragua, Costa Rica, Panama, and the northern portion of the east coast of Brazil. But not a single leader of any of these voyages of exploration had concluded that a new continent, which was located between Europe and Asia, had been discovered!

Only the two printed narratives that were related to the Vespucci voyages contributed to the conclusions of the members of the Gymnasium Vosagense. The earlier, *Mundus Novus*, initially published in 1503 and subsequently reprinted on several occasions, was certainly appreciated by the scholars at Saint Dié because one of the group, Matthias Ringmann, had edited a Latin edition of the book published in 1505. The influence that the second of the narratives, *Lettera*, had is attested to by the inclusion of a Latin translation of the full text in *Cosmographiae Introductio*.

Any conclusion regarding the influence of earlier maps—that depicted land within the Western Hemisphere—on the maps produced by Waldseemüller at Saint Dié in 1507 is purely conjectural and unsubstantiated. Even the starting point for a list of maps of America is difficult to define and open to argument. Among the wreckage of the *Santa Maria*, which foundered on a reef off Caracol Bay near the north shore of Hispaniola, a sketch map of a portion of that coast was found and has been attributed by some to Columbus himself. It includes four place names in addition to "la ispañola," "san nicolas," "tortuga," "monticristi," and "navida." Certainly, this document was unknown to the members of the Gymnasium Vosagense.

Another starting point might follow the listing and analysis of the distinguished historian of the cartography of America, Henry Harrisse. This brings into focus the two woodcuts that are included in the 1493 Basel edition of Columbus's letter to Luis de Santángel, which he wrote after his first voyage. The first edition was printed at Barcelona in April 1493 and contains no illustrations. The only known copy resides in the New York Public Library. The Basel edition, printed later that year by Bergmann de Olpe, is titled *Epistola de Insulis Inventis* and includes among its eight illustrations two that provide geographic information. One eight-by-eleven-centimeter woodcut depicts Columbus's ship and names "hyspana," "fernãda," "Ysabella," "Saluatorie," and "Conceptois marie" as a completely fanciful representation of the islands (see fig. 2A). Equally fanciful is the second illustration showing a multi-oared galley and a land named "Insula hyspana." (See fig. 2B.)

By the end of the fifteenth century, enough detailed information was gleaned to undoubtedly create several maps. The information was gathered from the first three voyages of Christopher Columbus that discovered islands in the Caribbean and the northern shore of Venezuela, John Cabot's discovery of Newfoundland, and Alonso de Hojeda's expedition to South America.

There is specific literary evidence that one map was made during Columbus's second voyage. The University of Bologna Library currently holds a manuscript letter from Michael de Cuneo, a gentleman who participated on that voyage, stating: "We saw many islands, which His Lordship the Admiral ordered to be inscribed on the map distinctly." A map by Alonso de Hojeda, providing a "graphic and minute description of the country discovered by Hojeda," is mentioned in the memorial addressed to Ferdinand and Isabella by Christopher Columbus from the city of Isabella, on January 30, 1494. According to letters included in *Sentencias Catholicas*, cosmographer/jeweler James Ferrer de Blanes sent a map to Ferdinand and Isabella on January 27, 1495. It was apparently made to illustrate Spanish claims related to the line of demarcation established at the Treaty of Tordesillas and consisted of a "description of the World, on a plane surface, in which can be seen the two hemispheres . . . and our Arctic [pole] and its opposite Antarctic [one]; the equinoctial circle, and the two tropics of declination of the sun." The phraseology related to "two hemispheres" is peculiar, because at that time, the concept of two hemispheres had not been promulgated. A map, which is known to have existed in Lisbon in 1847 and is said to have been used by Vasco da Gama on his voyage around the tip of Africa in 1497, supposedly depicted a large island in the New World, but remains lost.[24]

Several maps have been associated with the voyages of John Cabot. The dispatch of Raimondo Soncino, dated December 18, 1497, states: "This Mr. John has the description of the world on a map, and likewise on a solid globe which he has made, and he shows where he landed, and that sailing eastward [westward] he has passed

Figure 2A and B. Woodcuts from *De insulis in mari Indico nuper inventis,* Columbus's printed Letter Basel, 1493.

far beyond the country of the Tanais." This attests to Cabot's conclusion that he had discovered land, which continued westward and southward as part of the coast of Asia. Also related to Cabot's exploration, in 1498, Pedro de Ayala, one of the two Spanish envoys to the English court, wrote: "I have seen the map which the discoverer has made." The existence of that map was substantiated by Ruy Gonzalês de Puebla, the other leading Spanish ambassador, who related: "I have seen the route which they brought, and what they are in search of is [the country] which your Highnesses possess." None of these maps concerning Cabot's discovery has been uncovered.[25]

A map made by Christopher Columbus during the third voyage, on which crew members set foot on South American continental land, is referred to in several contemporary narratives. After the August 5, 1498, landing on the Paria Peninsula of Venezuela, Columbus wrote:

"Meanwhile I shall send to your Highnesses that description and the picture of the country." Alonso de Hojeda provides added evidence that a map was sent with his testimony: "I have seen the figure which the said Admiral at that time sent to Castile to the King and Queen, concerning what he had discovered." Bernardo de Ibarra also confirmed this when he wrote: "And he sent it describing with the said letter in a sea chart the rhumbs and winds by which he reached Paria." That map was said to have been copied and used by Alonso Niño, Alonso de Hojeda, and other navigators on their subsequent voyages. Francisco de Morales also wrote: "I have seen a nautical chart, which the Admiral made at Paria; and I believe that all have sailed by the same."[26]

In 1502, Bartolomé de Las Casas, the first ordained priest in the New World, went to Hispaniola. Around 1527, he began working on his lengthy tome, *Historia de las Indias* and continued writing it until his death at age eighty-five in 1566. The work is preserved in the Biblioteca National in Madrid and was very influential in its time.

But the text was not published until 1875. In that work, regarding the map made by Columbus during the third voyage, Las Casas indicates: "He sent also to the Kings the picture or representation of the country, which he had discovered, with the islands distinctly marked which laid adjacently, and, in writing, his entire voyage. By that picture or delineation of the land of Paria which he sent to the Kings, Alonso de Hojeda went there." Harrisse suggests that the map must have been drawn by someone other that Columbus, because he was suffering from ophthalmia at the time.[27]

The table of contents of a cartobibliography of maps of the Western Hemisphere is invariably headed by a map of the world by Juan de la Cosa (see plate 2). This extraordinary work—consisting of ink and watercolor artistry on oxhide, mounted on a background of Russian leather, and encased in a carved oak frame—is on display in the Museo Naval in Madrid. Entrance to the museum, which is located a short distance from the famous Prado art collection, is free, but a passport or acceptable identification is required because the facility is officially part of the Spanish navy.

This revered document itself has an extraordinary history that occurred during a time of increased interest in the accomplishments of Christopher Columbus. In 1825, Martin Fernández de Navarrette had published the first two volumes of *Collecion de los viages y descubrimentos que hicieron por mer los Español desde fines del siglo XV*, which, in turn, inspired the prolific American writer Washington Irving to write *The Life and Voyages of Christopher Columbus*, published in 1828, while Irving was serving as a diplomat in Madrid. Irving was the first to reveal that the explanation for the naming of America was to be found in *Cosmographiae Introductio* and, moreover, that the book specifically referred to an accompanying globe gore and also a flat map.

In 1832, at the time of a cholera epidemic, Baron Walckenaer, the Dutch ambassador to France, found the Juan de la Cosa map in a Paris "bric-a-brac," or curio shop. He recognized its potential importance and brought it to the attention of Alexander von Humboldt, the famous German naturalist and explorer, who authenticated the map after an extensive study. At the time, Humboldt was engaged in his monumental work *Examen Critique de l'Histoire de la Géographie des Nouveau Continent et des Progrès de l'Astronomie nautique aux Quinzième et Seizième Siécles*, and he made the map a main feature of this tome.[28]

After Baron Walckenaer died, the map was offered at auction in 1853.

The Bibliothèque Imperiale de Paris submitted one bid, and the distinguished bookdealer of Americana, Henry Stevens, offered 4,000 francs, the equivalent of $200 at the time. The successful bidder was Sr. Ramon de la Sagra, who purchased it for the Spanish government for the equivalent of $208. For the insignificant difference of $8, the first map of America remains in Spain, although Sagra indicated that he was prepared to bid the item up to any price.[29]

The map has evoked two major questions: Who was the mapmaker? and When was the map on display in the Museo Naval actually drawn? Most scholars identify the name of Juan de la Cosa, which appears on the parchment, as the Basque seaman and cartographer who sailed with Christopher Columbus on his first and second voyages. A Basque with the surname of El Viscaino, or Biscay man, one Juan de la Cosa, was the owner and master or mate of Columbus's flagship *Santa Maria*, and he accompanied Columbus on the second voyage. He sailed with Alonso de Hojeda and Amerigo Vespucci on the voyage to the Paria Peninsula in 1499, and with Rodrigo de Bastidas to the north coast of South America in 1500. In 1503, he commanded an expedition to Uraba (Panama) and between 1504 and 1506, he conducted a voyage of discovery to Darien (Panama). He sailed again to the Indies in 1507 and, in 1509, he settled with his family in Hispaniola. In November of that year, he accompanied Hojeda on an expedition to Darien, during which he was mortally wounded by a poisoned arrow. He died at Cartagena on February 28, 1510.

By contrast, Morison suggests that there were two Juan de la Cosas. The first was the one who commanded the *Santa Maria* on the first of Columbus's voyages and disgraced himself when his ship was grounded and wrecked. The other Juan de la Cosa, who conducted the subsequent voyages listed above and who died at Cartagena, was the cartographer that made the map in question.[30]

As far as the date of execution of the document on display at the Museo Naval, it is inscribed on the map "Juan de la cosa la fizo en el puerto de S; mjª en el año de, 1500" (Juan de la Cosa made it at the Port of Santa Maria in the year 1500). The date is under contention because several aspects of geography depicted in the Western Hemisphere were not defined until later. The map shows evidence of exploration of southern Brazil that did not take place until 1503. Also, Cuba is depicted as an island; the island was not circumnavigated until 1508 when Sebastián de Ocampo completed the voyage. In 1494, during Columbus's second voyage, Juan de la Cosa signed a statement with the rest of the crew that Juana (Cuba) was part of the Asian mainland.[31]

Many of the features within the Western Hemisphere offer testimony that the map was produced after Columbus's fourth voyage, which included sailing off the coast of Central America. A representation of St. Christopher is shown in the region where Columbus was on his quest for a westward passage during his fourth voyage. The shape of South America south of Cabo de la Vela is essentially cor-rect, and that had not been defined until after 1500. The map might have been made by Juan de la Cosa before his death, but the carto-graphic historian Nunn believes that it is a copy that was executed by someone else.[32]

The Juan de la Cosa mappemonde, measuring 180 × 960 cen-timeters (70 × 380 inches), is classified as a portolan, or sailing chart, after the style of maps drawn in Spain, Portugal, and Italy during the fourteenth and fifteenth centuries. The term is properly restricted to charts that give sailing directions and generally disregard the interior of landmasses, emphasizing coasts, currents, harbors, shoals, and winds. The charts also display a system of lines emanating from the center of compass roses. These are navigational aids that were first placed on the sailing charts in the Middle Ages to designate points of the compass.

The map is well preserved and in generally good condition, although it is missing several small pieces including a two-inch-wide piece along the northern coast of Brazil. The map, which depicts a cir-cumference of 360°, incorporates Columbus's erroneously small esti-mation of the earth's diameter. The landmass in the Western Hemi-sphere is disproportionately large when compared to Europe, Africa, and the Mediterranean Sea. The absence of latitude and longitude and the style of projection preclude easy identification of specific areas. Many islands are defined in the Caribbean Sea. Cuba is depicted as an island with a fishhook shape and is named as such for the first time on any map. Columbus did not regard it to be an island but rather part of a Chinese province and named the island "Juana" for Queen Isabella's daughter, Doña Juana Infanta. The chief interest of the map is its depiction of the east coast of continental land in the Western Hemi-sphere, although it was assumed that the land was a part of Asia. Off the coast of Brazil, "Ysla descubierta por portugal" (Island discovered by Portugal) represents an interpretation of the land discovered by Cabral in 1500. "Este cavo se descubrio en año de mil y IIII XCIX por Castilla syendo descubridor vicentians" (This cape was discovered in the year 1499 for Castile, Vicente Yáñez being the discoverer thereof), referring to Pinzón's discovery in January 1500 (new-style calendar). "Mar Dulce" refers to the fresh water of the mouth of the Orinoco River seen by Juan de la Cosa during Columbus's second voyage. "Costa de perlas" (Pearl Coast) on the northern coast of South America was discovered by Columbus on his third voyage in 1498 and visited later by la Cosa sailing with Hojeda in 1499.

Placed along the northeastern coastline of North America are five English standards. The phrases "mar descubierta por yngleses" (sea discovered by the English) and "cavo de ynglaterra" (cape of England) offer irrefutable evidence of John Cabot's exploration in 1497. It is believed that la Cosa derived this delineation from a copy of Cabot's map (now lost) that was sent to King Ferdinand by Pedro de Ayala, the Spanish ambassador to England. The North American continent displays twenty inscriptions, including seven capes, a river, an island, and a lake. The names are not significant, since they are not found on subsequent maps.[33]

Obscuring Central America is a vignette showing St. Christopher, with a pine tree as a staff, carrying the infant Jesus over water, perhaps representing allegorically Christopher Columbus crossing an ocean to spread Christianity to the natives. Biblical reference also appears in the form of the three wise men bearing gifts across Asia. A tower is seen at Babylon and "R. Got" and "R. Magot" suggest Gog and Magog.[34]

The Juan de la Cosa map is generally presented as the oldest extant map based on the transatlantic discoveries of the Europeans. It apparently was not available to other contemporary cartographers and had little influence on the subsequent mapping of the Western Hemisphere or, more specifically, on the production of the 1507 world map by Martin Waldseemüller.

Other manuscript maps produced at the onset of the sixteenth century are relevant. In the *Probauzas del Fiscal* manuscript in the Archiva de las Indias, a declaration states: "Pedro de Ledesma said that he saw Vicente Yañez [Pinzón] and his companions sail for the voyage mentioned in the question [Pinzón's 1499–1500 voyage], and that the said

witness saw him return and bring a map of all he had discovered; and
that this map was inserted in the *Padron* of His Highness." Also, in the
same manuscript, it is written: "Pedro de Ledesma saith that when Diego
de Lepe sailed on his voyage of discovery, he saw the ships and men
depart, and return to Seville, except that the said Diego de Lepe who did
not come; and that those who had been with him brought a map of their
discoveries, which marked that its was from said point [*viz.*, The Cape of
St. Augustine], as far as the coast which trends southward."[35]

Arias Perez Pinzón also asserted that a map was shown to him by
Velez de Mendoza on his return to Spain. In addition, Peter Martyr
d'Anghiera indicated that, in 1513, he saw a Portuguese map that was
owned by Bishop Juan de Fonseca of Seville and that had been made
by "Americus Vesputius." Reference is made to Uraba in what is now
Panama. Peter Martyr writes in this "Second Decade" that was
addressed to Pope Leo X:

> This continent extends into the sea exactly like Italy, but is dissim-
> ilar in that it is not the shape of a human leg. Moreover, why shall
> we compare a pigmy with a giant? That part of the continent begin-
> ning at this eastern point lying toward Atlas, which the Spaniards
> have explored, is at least eight times larger than Italy; and its
> western coast has not yet been discovered. Your Holiness may wish
> to know upon what my estimation of *eight times* is based. From the
> outset, when I resolved to obey your commands and to write a report
> of these events, in Latin (though myself no Latinist), I adopted pre-
> cautions to avoid stating anything which was not fully investigated.
>
> I addressed myself to the Bishop of Burgos, whom I have already
> mentioned and to whom all navigators report. Seated in his room, we
> examined numerous reports of all those expeditions, and we likewise
> studied the terrestrial globe on which the discoveries are indicated
> and also many parchments, called by the explorers, "navigators'
> charts." One of these maps had been drawn by the Portuguese, and it
> is claimed that Americus Vespucci of Florence, assisted in its compo-
> sition. He is very skilled in this art, and has himself gone many
> degrees beyond the equinoctial line, sailing in the service and at the

expense of the Portuguese. According to this chart, we found the continent was larger than the caciques of Urabá told our compatriots.[36]

If it was a map made by Amerigo Vespucci during his Portuguese voyages, it would have been executed in 1502 when he returned from his voyage to South America under the Portuguese flag.[37]

A map is said to have been made at Palos, Spain, for the Venetian admiral Domenico Malipiero, based on data provided by Christopher Columbus from the first three voyages. Arias Perez Pinzón referred to another map made by Rodrigo de Bastidas and Juan de la Cosa delineating their discoveries on the north coast of South America. Narratives indicate that Columbus had access to marine charts of other mariners who had sailed along the so-called Pearl Coast of South America and that he suppressed a map of Darien (Panama) drawn by the sailor Pedro Mateos, who had viewed that area. None of the maps drawn subsequent to the Juan de la Cosa mappemonde of 1500 has been discovered, and, certainly, they did not directly influence the geographic conclusions expressed on the 1507 Waldseemüller world map.[38]

The "Cantino Planisphere" of 1502 (see plate 3) generally follows the Juan de la Cosa map in the chronology of the mapping of America, but some historians are of the opinion that it deserves the accolade of primacy because its date has been authenticated. Most historians, however, believe that the western portion of the Juan de la Cosa map was not drawn in 1500—the date on the map—but rather sometime after 1502 and probably closer to 1510. The Cantino map, a beautiful pen-and-ink drawing with added color and gilt, on vellum, and measuring 110 × 220 centimeters (43 × 86 inches), like the Juan de la Cosa map, has its own dramatic history.

Albert Cantino was the envoy representing Ercole d'Este, Duke of Ferrara at the court of Portugal in October 1501, when the two remaining ships from Gaspar Corte-Real's fleet of three returned to Lisbon from a voyage in the northern part of the Western Hemisphere without their commodore. In a letter to the duke, dated October 17, 1501, Cantino informed his employer of the success of the expedition and its exciting discoveries. Because the Portuguese king had placed an embargo on all charts displaying new discoveries, Cantino ordered a map that was made by a cartographer in Lisbon, requiring ten months between December 1501 and October 1502 to complete, at a cost of twelve gold ducats. The chart was clandestinely carried by Cantino to Genoa and then sent on to the House of Este, in Ferrara, where it remained until 1592.[39]

That year, Pope Clement VIII relieved Cesare d'Este of his duchy, and the entire ducal collection was moved to Modena. At some point, the top of the map, bearing the title in large Gothic letters, was cut away to accommodate its use as a covering for a screen. When the palace was looted by republicans in 1859, the map was stolen, but it was discovered in 1870 by Signor Boni, librarian of the Biblioteca Estense. He noticed it in a butcher shop on the Via Farini, still serving its purpose as a screen cover. He bought the map, removed it from the screen, and presented it to his library where it still resides.[40]

The map is oriented politically to the Portuguese point of view. It ignores Cabot's discoveries and delineates land west of Greenland but east of the papal line of demarcation, and, therefore, within the Portuguese realm. The land is adorned with the Portuguese flag and a legend stating: (translated) "This land was discovered by order of the Most High and Excellent Prince King Dom Manuel of Portugal. It was found by Gaspar Corte-Real, one of his noblemen, who, upon discovering it, sent [thence] a vessel with men and women of that country. He remained with the other vessel, but never returned [home], and the belief is that he was lost. The country contains much mast timber."[41]

The east and south coasts of Newfoundland from the Strait of Belle Isle to Placentia include a series of capes and bays. This suggests that Gaspar Corte-Real sent home a chart of the region, while the west coast is shown without any definition because it was not charted. A series of islands appear in the Caribbean sea, and Cuba has the same shape as on the Juan de la Cosa map. To the west of that island there is a peninsula—named "Isabella" on the chart—that Harrisse believes is the first representation of Florida. He states, "The critical historian of maritime discovery is justified in considering the north-western delineations in the 'Cantino Planisphere' as representing a continental region really existing. Now what is that country? Necessarily a portion of the Atlantic shores of the present United States, shown to have been discovered, visited, named, and described so far back as the close of the fifteenth century." Another map historian, Edward Stevenson, concurs. Others suggest that the mapmaker duplicated Cuba first as an island, and then, in accordance with Columbus's concept, as part of the mainland. Yet another explanation is that "Isabella" is Cuba and the peninsula is part of the Asian mainland that both Columbus and Cabot thought they had reached.[42]

The Brazilian coast also displays Portuguese flags and a legend reporting the landing of Cabral in 1500. The entire east coast of South America and Newfoundland are placed to the east (within the Portuguese domain) of the papal line of demarcation as stipulated in the 1494 Treaty of Tordesillas, which divided the New World between Spain and Portugal. The line, shown on a map for the first time, prominently courses its entire length.

In 1890, the discovery of an undated manuscript map, which was found in the Archives du Service Hydrographique de la Marine in Paris, was announced. Measuring 115 × 225 centimeters (45 × 89 inches), it is a manuscript most likely drawn in Portugal in 1504–1505 by a Genoese whose name was initially thought to be Nicolay de Canerio but was recently reinterpreted as Nicolo Caveri (see plate 4). The geographic configurations and nomenclature are similar to the Cantino map, but a new feature is the inclusion of a scale of latitude. The southern continent extends from 12° to 35° south latitude, about 10° more than the Cantino map. On the east coast of Brazil, "Porto de. Sto. Sebastiano" and "Alapago de San Paullo" appear for the first time. It is believed that the map incorporates the land discovered by Columbus during his third voyage.[43] On the Brazilian coast, a legend states:

The True Cross, so called, which was discovered by Pedro Alvarez Cabral, gentleman of the household of the King of Portugal; and he discovered it in navigating as chief captain of fourteen ships which the said King was sending to Calicut, and in following his route, he found that land, which is believed to be a continent, where are many men endowed with reason, and men and women who go naked, as brought into the world. They are rather white than dark, with smooth hair. The land was discovered in the year [one thousand five hundred].[44]

The northern continental region extends from 50° to 20° north latitude. The name "Lago de lardo" appears for the first time and is considered to be the antecedent of "larro dellodro" on the Waldseemüller 1507 world map. Newfoundland has the same island configuration as on the Cantino map and is located in the same incorrect latitude, containing no name and unconnected with the landmass, which questionably represents the Florida peninsula. The papal line of demarcation is not shown. Although the depiction of Cuba, Yucatán, and Florida are only partially correct, this portrayal was replicated on maps for more than twenty years.[45] Most striking is the presentation of a delineation of a Gulf of Mexico flanked by the peninsulas of Florida and Yucatán. Over the West Indies, a legend states: "The Antilles of the King of Castile, discovered by Collonbo, Genoese [this word is not in Cantino] Admiral, which islands were discovered by command of the very high and very powerful prince the King Dom Fernando, King of Castile."[46]

In their analysis, professors Joseph Fischer and Franz Ritter von Wieser conclude "that Waldseemüller, when elaborating his map of 1507, had before him not only a chart of the Canerio [Caveri] type, but the Canerio chart itself." "A small island east of 'Spagnolla' on the Waldseemüller World Map of 1507 bears the queer designation: *Laonizes mil virginum* just as on the Canerio chart *Laonizes mil virgines*, instead of: *Las omze mjll vigines* as it is correctly given on the Cantino chart." "An island east of 'Isabella' with Juan de la Cosa bears the name: *maiuana*; with Cantino: *ilha managua*; with Canerio *mag^{na}na*; with Waldseemüller on the world-map of 1507: *magna*, the superscribed syllable 'na' having evidently been overlooked. On the

'Tabula terre nove' of the Strassburg Ptolemy of 1513 and on the 'Carta marina' of 1516 Waldseemüller correctly reads *magnana* after Canerio." Edward Stevenson agrees with Joseph Fischer and indicates that "[s]ince the finding of the long-lost Waldseemüller maps by Professor Fischer we are able to trace the influence of the great German cartographer through a line of important maps of the sixteenth century. As there can be little doubt that Canerio [Caveri] was his teacher, through his chart of 1502, it is fitting at this time to honor that Italian chart-maker with this carefully executed reproduction of his great work."[47]

A map depicting the fourth voyage of Christopher Columbus was drawn by Bartholomew Columbus at the completion of that voyage in November 1504 and, because it is known that Bartholomew went to Italy in 1506, it had to be completed before that time. Peter Martyr, the historian in King Ferdinand's court, reported that he saw the map in 1513, when it was possessed by Bishop Fonseca. A copy of that map, probably drawn by Allessandro Zorzi sometime between 1516 and 1522, was discovered at the Bibliothèque Nationale in Florence by Professor von Wieser. That was about a decade before he collaborated with Professor Fischer on work related to the 1507 Waldseemüller world map.[48]

Bartholomew Columbus's map consisting of three sketches encompasses the entire world in the equatorial region (see plate 5). On the first of the three primitive sketches are depicted the places in Central America that Columbus visited during his fourth voyages. Columbus's conclusion that the continental land he discovered was part of Asia is manifest on the map, graphically amplifying a letter Christopher Columbus sent to King Ferdinand from Jamaica on July 7, 1503. The names appearing on the coast west of the West Indies— "cariai," "carambaru," "bastimentos," "retrete," and "belporto"—on the first sketch are also shown on the third sketch along the Asian coastline. The distance between Europe and Asia is markedly underestimated, in accordance with Columbus's calculations.[49]

On the first sketch, the landmass in the northwest, bordering

what could be interpreted as a Gulf of Mexico, is unequivocally con-
nected to the Asian mainland. To the south, a narrow isthmus in the
region of current Panama is shown, as is a strait emptying into the
"Sinus Magnus," the classical name for the waters west of Asia. South
America, with "Paria" included along its north coast, is named
"Mondo Novo" (New World), a term erroneously credited to
Amerigo Vespucci's publication *Mundus Novus*.

Only one potentially influential printed map might have been
published before the Waldseemüller world map. The only known copy
is preserved in the British Library and is included for completeness
because it bears the date of 1506. It has been assigned priority as the
oldest known printed map to show part of America. The map (see
plate 6) is referred to as the Contarini-Rosselli world map, crediting
the cartographer, Giovanni Matteo Contarini, and the engraver,
Francesco Rosselli. Bookdealer and bibliophile Henry N. Stevens, who
has connections with the Waldseemüller map of 1507 on many levels,
concludes that if the map were printed in Florence, as is generally
accepted, the date of 1506 on the map could have referred to any time
between March 25, 1506, and March 24, 1507, on the new calendar.

North America is presented as an extension of Asia. In the ocean,
off the most eastern tip—in the region of current Newfoundland—is
an inscription stating, "This land the seamen of the King of Portugal
discovered," which refers to the explorations of the Corte-Reals.
South America bears the name "Terr. S. Crucis," and off the southeast
coast a legend reads: "This is the land named Santa Cruz which was
lately [discovered] by the most noble lord Pedro Alvarez [Cabral] of
the illustrious stock of the most serene King of Portugal in 1499." On
the north coast of the southern continent is the statement: "This is
the gulf in which the Spaniards found very many pearls, and along
this coast lions, swine, stags, and other kinds of animals."[50]

The islands of "Terra de Cuba" and "Insula Hispaniola" appear,
accompanied by the legend: "These are islands which Master Christopher Columbus discovered at the instance of the most serene King of
Spain." Columbus's persistent belief that he had reached the coast of

Asia is manifest in a legend that appears off the east coast of that continent: "Christopher Columbus, Viceroy of Spain, sailing westwards, reached the Spanish islands after many hardships and dangers. Weighing anchor then he sailed to the province called Ciamba. Afterwards he betook himself to this place, that most diligent investigator of maritime things, asserts, holds a great store of gold."[51]

It is likely that the Contarini-Rosselli map actually postdated the creation, if not the printing, of the 1507 Waldseemüller world map. It could not have influenced the scholars at the Gymnasium Vosagense. The answer to the question as to what did influence the ultimate expressions, which were presented in the text and on the two maps emanating from that scholarly group, remains conjectural.[52]

As the revolutionary geography that is presented on the Waldseemüller world map of 1507 is reassessed at the beginning of the twenty-first century, questions abound. Why does the Waldseemüller world map depict two continental landmasses in the Western Hemisphere interposed between two great oceans, ostensibly the Atlantic and Pacific? Why are the continents separated by a strait on the main map while the inset of the Western Hemisphere shows an isthmus joining North America and South America? How did a western coast of each of the continents come to be shown six years before Balboa supposedly became the first European to view the Pacific Ocean from the Darien Peninsula (Panama) and twelve years before Magellan's voyage of circumnavigation, during which the straits at the southern tip of South America were traversed? Why, on the inset map of the Western Hemisphere, is the west coast of both the northern and southern continents depicted by a series of straight lines, and why is the west coast of North America placed along a meridian? By contrast, on the main map, why are mountains shown along the entire west coast of North America to the west of a curved, uninterrupted line seeming to define the western limits of "Unknown Land"? And on the southern coast of South America, as represented on the main map, why is the northern portion shown as a simple uninterrupted line bordering "Unknown Land," while the southern portion includes a west coast characterized by irregularities suggestive of many capes and bays? On what basis was a bulge at the junction of the current Chile and Peru, extending into the Pacific Ocean, positioned in its essentially correct location? Why is a mountain range east of the west coast of the southern continent in the actual location of the Andes?

Although not one of these many questions can be answered based on irrefutable evidence, speculation continues to pour forth. In a lecture given at the Library of Congress in October 2002, Peter W. Dickinson, a political-military analyst at the CIA and an amateur geographer and scholar of the Age of Discovery, related his theory. Dickinson concluded that many of the questions raised by the geography presented on the Waldseemüller map can be explained by accepting the premise that some European explorers, probably Portuguese, had traversed the Straits of Magellan and sailed as far north as Acapulco, Mexico, before 1506. He further asserted that the findings of these explorers had been drawn on charts that Waldseemüller either had access to or that he was informed of their representations. For example, the placement of a mountain range on the western edge of South America, south of the equator, in the location of the Andes, suggests a viewing of these impressive high peaks and rises of land from the sea because there is no evidence of overland expeditions from the east coast before the 1507 date of the map.[53]

Reinforcing this conclusion is the relatively correct shape of the western coastlines of South America and Central America, as depicted on the inset map of the Western Hemisphere. On both the inset map and the main map, there is a sharp shift to the northwest in the western coastline of South America where Peru and Chile meet. As a consequence, the Waldseemüller map captures the essence of the shape of South America. A comparison of widths of South America, as calculated on the inset and main map at the equator and at intervals of 10° to 40° south latitude, shows a remarkable agreement with the true width. Some researchers believe that this could only be explained by mariners having sailed along the west coast of the continent from the south to as far north as Acapulco, taking astronomical measurements along the way and adding these findings to those previously made during the known early sixteenth-century voyages along the east coast.

The third major element in support of Dickinson's theory relates to the name "Oceanus Occidentalis" (Western Ocean, i.e., the Atlantic Ocean) that appears to the west of the southern continent in the Western Hemisphere on Waldseemüller's globe gores, which were part of the 1507 publication of the Gymnasium Vosagense as indicated in the *Cosmographiae Introductio*. This depiction of the ocean suggests that the cartographer was expressing the fact that the sea west of South America was an extension of the Atlantic Ocean. The explorers prior to 1507 had discounted finding a passage to the west of Cuba, and the 1500 Juan de la Cosa map depicts no strait in the Central American coastline. Dickinson postulates that the confluence of waters east and west of South America had been defined by a passage that predated Magellan's—through the later-to-be-named Straits of Magellan.

The existence of a Portuguese chart or charts from which Waldseemüller derived his information is reinforced by a passage in the published account of Magellan's voyage by Antonio Pigafetta, who was a participant. Therein it is stated: "The strait was a round place surrounded by mountains, as I have said, the greater number of sailors thought that there was no place by which to go out thence to enter into the peaceful sea. But the Captain-General [Magellan] said that there was another strait leading out; and that he knew well, because he had seen it [on] a marine chart of the King of Portugal, which map had been made by a great pilot and mariner named Martin of Bohemia."[54] Martin Behaim was a noted cartographer, famous for a globe he made in 1492 just before Columbus's discovery. Behaim settled in Portugal, and it is possible that Magellan, while serving as a clerk in the Casa de India, saw a chart made by Behaim.

The impact of prior Portuguese maps on the Waldseemüller world map and the suggestion that information had been gathered concerning the west coast of the continent or continents in the New

World before the map was produced serve as a springboard for another theory: that the Chinese discovered America in the early part of the fifteenth century and provided the geographic information on which the pre-Columbian Portuguese maps were based. Gavin Menzies, a British submariner and amateur historian, has devoted his efforts to providing evidence that the great Chinese fleet sailing from 1421 to 1423 encountered Australia, New Zealand, South America, Central America, and both coasts of North America.[55]

It is known that in March 1421, Emperor Zhu Di dispatched a huge armada, consisting of four great fleets led by Admiral Zheng He, with the charge to "proceed all the way to the end of the earth to collect tribute from the barbarians beyond the seas . . . to attract all under heaven to be civilized in Confucian harmony."[56] The elements of the fleet that survived returned in October 1423. What ensued in the interim, what discoveries were made, and what lands were set foot upon remain unknown, because when the ships returned all records were deliberately destroyed by a new emperor. This was part of the new emperor's policy to return China to an agricultural economy characterized by xenophobia and isolationism.

In reference to the Western Hemisphere, Menzies concluded that one of the fleets rounded the Cape of Good Hope, sailed westward to Patagonia, through the Straits of Magellan and to the South Shetland Islands north of Antarctica, and then eastward to Australia. Another fleet sailed up the west coast of South America as far north as Ecuador before crossing the Pacific Ocean on a journey to Australia. One fleet sailed east from the Philippines to the west coast of North America and entered San Francisco Bay.

All of these voyages are, of course, speculative, emanating from the mind of Gavin Menzies, and are based solely on deductive reasoning using diverse evidence, none of which has been authenticated. The age of a Chinese wooden pulley found in the northwestern region of the United States, a part of a Chinese junk near Sacramento, and an anchor in Ecuador have not been established. The origin of Asiatic chickens in the Americas and of maize in China, used as evidence for

Menzies' thesis, has not been resolved. As yet another part of Menzies' deductive speculation, part of the Chinese fleet supposedly journeyed through the Caribbean and then along the east coast of Florida, continuing as far north as Rhode Island.

In creating his broad scenario, Menzies invokes evidence from several pre-Columbian and pre-Waldseemüllerian maps. One map was reported by Portuguese historian Antonio Galvão to have been brought back from Venice by Dom Pedro to Henry the Navigator in 1428. The translated narrative states: "from whence he brought a map of the world, which had all the parts of the world and earth described. The Streight of Magelan was called in it the dragons taile."[57] This is offered by Menzies as an assertion that the "Streight of Magelan" had been charted years before Magellan's sixteenth-century voyage of circumnavigation. The map has been lost, but Menzies believes that Christopher Columbus possessed a copy of the map and that the information depicted on that map was also incorporated in the Piri Resi map of 1513. In 1431, Henry the Navigator sent Gonzalo Velho Cabral in search of the islands that are supposedly shown on the 1428 map.[58]

Another map invoked by Menzies in defense of his thesis is a chart dated 1424, signed by the Venetian cartographer Zuane Pizzigano. The manuscript was discovered about seventy years ago and purchased by the James Ford Bell Collection at the University of Minnesota in the early 1950s. Armando Cortés states: "The great importance of this chart lies in the fact that it is the first to represent a group of islands in the western Atlantic called Saya, Satanazes, Antilia, and Ymana. . . . [T]here are many good reasons for concluding that the Antilia group of four islands shown for the first time in the 1424 chart should be regarded as the earliest cartographic representation of any American lands." Menzies contends that "Satanazes" is Guadeloupe; "Antilia" is Puerto Rico; "Saya" is Les Saintes; and "Ymana" became Rossellia. He believes that the fruit found by Columbus's crew on these islands had been brought there by the Chinese fleet in 1421–1423.[59]

Menzies also postulates that the voyage of Cabral in 1431 resulted in a Portuguese settlement in Puerto Rico at that time, sixty-one years before Columbus arrived. According to Antonio Galvão, a ship arrived in 1447 through the Strait of Gibraltar with men who came from the Antilles or New Spain.[60]

The 1507 world map by Waldseemüller has a prominent role in the support structure of Menzies' conclusions. He contends that the latitudes on the Waldseemüller map correspond to those of Vancouver Island in Canada down to Ecuador. The gap between the two continents of the main map is explained by the fact that the Chinese sailed into the Gulf of Tehuantepec in Guatemala and, unable to continue westward because of shallow water, turned back, concluding that the shallow water joined the two great oceans. Zooming in on the Waldseemüller map with what some consider great imagination, Menzies identifies representations of the Brazos, Alabama, Roanoke, Delaware, and Hudson rivers, in addition to almost one thousand miles of the Mississippi River.[61]

Until substantiated by factual evidence and enhanced by appropriate analysis, Menzies' conclusions remain a single individual's broad speculative opinion. As such, they can hardly be used to explain the enigmas of the geography presented on the Waldseemüller map. Hans Wolff, referring to the west coast and southern part of South America in his 1992 book *America: Early Maps of the New World*, stated: "It is not clear whether this accuracy was due to lost or secret knowledge, or was, in fact, coincidence or intuition." Currently, almost all scholars believe that Waldseemüller's 1507 depiction of the western coastlines of the two continents represents chance and artistic license.[62] According to Johnson: "America's status as a separate island was thus a tidy solution to the untidy mess of tantalizing, yet incomplete, information that German cosmographers had been handed. In showing these lands as distinct from Asia, Waldseemüller and Ringmann were not so much relying on the claims of Vespucci as making their own adjustments in order to weave together ancient categories and new knowledge."[63]

There is no mention of Portuguese source material in the text of the *Cosmographiae Introductio*. Waldseemüller's acquaintance with either a manuscript chart or a narrative defining the interposition of the two continents in the Western Hemisphere, between two great oceans, is negated by his subsequent maps published in the 1513 Ptolemy *Geographia*. Both the "Tabula Terre Nova" and the "Orbis Typus Universalis" depict only the east coast of the lower portion of North America and the southern portion of South America. The same situation pertains to Waldseemüller's "Carta Marina" of 1516.

Most recently, by applying advanced mathematical techniques to the Waldseemüller world map of 1507, a cartometric study of the coast of South America was conducted by John Hessler. Polynomial-warping algorithms and polynomial regression analysis was applied to the map after it was digitized. By measuring the distance between one hundred corresponding points on the east and west coasts of South America along lines of constant latitude, the width of the continent on the map was determined.[64]

The text of *Cosmographiae Introductio* suggests that the cartographer drew the map based on geographic evidence rather than speculation. In the book, it is written: "Hunc in modu terra iam quadripartite cognscitiet: et sunt tress prime partes continentes / Quarta est insula cu omni quaque mari circudata conspiciat." ("The earth is now known to be divided into three parts. The first three parts are continents, while the fourth part is an island, because it has been found to be surrounded on all sides by water.")

The use of the terms "now known" and "has been found" suggests prior knowledge on the part of the cartographer. The mathematical analysis substantiates this and provides confirmation that Waldseemüller knew that value of distances between points along latitudinal lines. The accuracy displayed by distances between points on the east and west coasts of South America is highly suggestive of a factual basis for the depicted geography.[65] The basis of the geography depicted of Waldseemüller's map remains an unsolved mystery.

If we return to the first decade of the sixteenth century, it is possible to envision a small group of long-robed scholarly men in Saint Dié preparing to disseminate secular knowledge with the aid of a newly acquired printing press. Before setting the movable type, they had to make decisions. They adopted a theme. Recent changes in appreciation of the world's geography, with an emphasis on discoveries in what had been referred to as the New World, merited emphasis. Input from the cumulative knowledge of the savant participants ensured the inclusion of all pertinent material. Communal sifting separated facts from fiction. Books and individual documents, including manuscripts, contributed to a final assessment. A small book with the revolutionary title *Mundus Novus*, which had been edited previously by one of the scholars, and a recently acquired report of the voyages of Amerigo Vespucci captured the interest of the scholars and dominated the basis for their conclusions.

Doubtless, the members of the group with specific preparation in geography and, more particularly, cartography were charged with the responsibility of providing graphic accompaniments to the narrative. Because one of the originally asserted goals of the scholars was to produce a new Ptolemaic atlas, the creation of pertinent maps that would depict the New World was a prime focus. Some previously prepared manuscript maps may have been viewed personally. Perhaps some of these maps depicted Portuguese voyages, the reports of which were suppressed for political reasons, including: the strait later named for Magellan, the west coast of the American continents, and an insular concept for the continental landmass in the New World. In the case of other maps, descriptions of their representations were analyzed. Based on these direct and indirect evaluations, conclusions were reached, and a pictorial document was synthesized in order to emphasize a *new* geography.

The period of preparation came to an end, and the time arrived for finalization. The new press at the small, walled community of Saint Dié issued its first production in April 1507, a book titled *Cosmographiae Introductio*, which specifically announced the production of two new maps.

CHAPTER 3

THE NARRATIVE
AND ITS NUANCES

Books are the treasured wealth of the world and the fit inheritance of generations and nations. . . . Their authors are a natural and irresistible aristocracy in every society, and more than kings and emperors effect an influence on mankind.

Henry David Thoreau

In the early years of the sixteenth century, the scholars who had assembled at the Gymnasium Vosagense in Saint Dié were focusing on distilling recently acquired information and disseminating knowledge in the realm of cosmography. This included the geography and astronomy of the known universe. Their plan was to produce maps for a new atlas that would for the first time incorporate a geographic representation of the New World in the Western Hemisphere. No previously published atlas had included such material. Although maps were the focus of the project, it was logical and expeditious to produce a text that would explain the basis of a revolutionary new cartographic representation of the world. The vehicle for transporting the verbal and pictorial information was to be a mechanical device specifically acquired by the Gymnasium Vosagense to facilitate the process—the printing press. The press available in Saint Dié was capable of producing a small book but was not sufficiently large enough to print the sheets required for the production of a large map.

The characteristics of the recently developed mechanical device had evolved in Mainz, Germany, only a half century earlier in the middle of the fifteenth century. The process of what was at the time considered to be modern printing incorporated several interrelated elements. Metal type of separable letters with an even top surface when assembled were made of an alloy of antimony and bismuth, blended with lead and tin. The mixture expanded when passing from a molten to solid state. The casting process produced individual letters with dimensions that were identical in height and width. This allowed for adjustment for the varying widths of the letters. The characters had to be capable of reuse and redistribution to structure new words.

A second essential element was an oil-varnish ink that adhered to the metal type. Another requisite was a press that allowed for consistent transfer of the ink from type to paper. The first product of the process of modern printing was the forty-two-line Bible, an elegant folio-sized book containing 1,286 pages in two volumes that came off the press of Johannes Gutenberg around 1455.

In keeping with the focus on cosmography, which was entertained by the scholars of the Gymnasium Vosagense, the Saint Dié press—a modest derivative of the Gutenberg press—issued its first publication, *Cosmographiae Introductio*, on April 25, 1507, a date that appears with the book's colophon (fig. 3) as VII Calends May. "The book is a small quarto (about 240 × 138 mm) (9.5 × 5.5 in) of 52 leaves (104 pages), spaced for 27 lines to the page, made up with roman type imprinted on paper bearing two watermarks, a bull's head with a clover leaf at the top of a staff that rises between the horns, and a large five pointed star." It includes four figures and a foldout plate of woodcuts.[1] Four distinct variants of the book have been identified; two with the colophon dated VII Calends May (April 25), and two dated III Calends September (August 29). American historian John Boyd Thacher

Figure 3. Colophon, *Cosmographiae Introductio.*
From Charles George, *The Cosmographiae Introductio*
of Martin Waldseemüller in Facsimile
(New York: United States Catholic Historical Society, 1907).
Courtesy of Herbermann.

points out that the letters "M. I.," designating Martin Ilacomilus (Waldseemüller), are significantly larger than the initials "G. L." and "N. L." for Gaultier Lud and Nicolas Lud. This is offered as evidence that the Gymnasium Vosagense recognized the dominant role of Martin Waldseemüller in the production of the work.

The copy of *Cosmographiae Introductio* in the Lenox Library of the New York Public Library is generally referred to as the earliest of the four variants. Like some of the manuscript maps described in the previous chapter, that copy also has an intriguing history. During the first half of the nineteenth century, the respected French geographer Jean Baptiste Eyrièz purchased the map for the equivalent of twenty cents from a dealer in a secondhand bookstall along the Seine River in Paris. It was shown to Alexander von Humboldt, who discussed its importance in the fourth volume of his *Examen Critique*. Washington Irving, whose "Legend of Sleepy Hollow" and "Rip Van Winkle" had already received international acclaim, was the first to trace the earliest appearance of the name "America" to the *Cosmographiae Introductio*.

In 1828, Irving wrote in *The Life and Voyages of Christopher Columbus*: "The first suggestion of the name appears to have been in the Latin work already cited, published in St. Diez [sic], in Lorraine, in 1507, . . . The author after speaking of the other three parts of the world, Asia, Africa, and Europe, recommends that the fourth shall be called Amerigo, or America, after Vespucci whom he imagined to be its discover."[2]

After the death of Eyrièz, a recognized collector from Lyons, Nicolas Yéméniz, purchased the book for the equivalent of thirty-two dollars. The book was next bought at auction by Almon W. Griswold of New York in 1867. After it temporarily resided in the collection of Henry C. Murphy of Brooklyn, it was procured by the Lenox Library, which was subsequently incorporated into the New York Public Library, where it is now displayed in an elegant binding created by Trautz-Bauzonnet as one of the distinguished library's main treasures.[3] Marie Armand d'Avezac distinguished the four variants by the first line of the title page and the date of the colophon:[4]

Edition I	Cosmographiae Introdu	vij kl' Maij
Edition II	Cosmographiae Introductio	vij kl' Maij
Edition III	Cosmographiae	iiij kl' Septembris
Edition IV	Cosmographiae Introdu	iiij kl' Septembris

Wilberforce Eames, librarian of the Lenox Library, created a more detailed analysis of the variations of the four editions. American historian John Boyd Thacher disputed the priority presented for the four editions by d'Avezac and Eames, and supplied evidence to establish a different sequence of publication of the *Cosmographiae Introductio*, suggesting that Edition II actually preceded Edition I.[5] Thacher concludes that the so-called Edition I is a mongrel and does not merit the status of priority. But, because April 25, 1507, is the date of both variants, April 25, 1507, remains the date on which the word "America" first appeared in print.

Henry C. Murphy—who owned the so-called Edition I and had the opportunity to examine all four variations of the *Cosmographiae Introductio* printed at Saint Dié—affirms Thacher's conclusion. E. D. Church, by contrast, disagrees with Thacher and Murphy, and endorses the d'Avezac assignment of priority for the variations. Joseph Sabin and associates define two editions printed at Saint Dié on fifty-two leaves on April 25, 1507, two Saint Dié editions with the same number of pages printed on August 29 of the same year, an edition printed on thirty-two leaves at Strasbourg by Grüninger in 1509, and an edition printed at Lyons in 1517–1518 on thirty-three leaves. The number of printings over an extended period of time and the identification of thirty-one extant copies indicates that the *Cosmographiae Introductio* evoked significant interest. Ferdinand Columbus—Christopher's son—owned an annotated copy that has been preserved.[6] The *Washington Post* reported on November 29, 1924, that the village of Saint Dié had recently purchased a copy of the book from its neighbor, Nancy, for twenty-eight thousand francs, thereby allowing the "godmother" to regain the birth certificate of the "god child."

The question as to why there were essentially two parallel printings of the work from the Saint Dié press in 1507 is intriguing. One of the printings, consisting of two editions, contains a dedication offered in the name of the gymnasium, while the other, also represented by two editions, by contrast, contains an address by Matthias Ringmann and a dedication by Waldseemüller. In the latter two editions, on the verso of the title page, a ten-line poem is addressed to Emperor Maximilian by Philesius of the Vosges (Ringmann). Beginning on the recto of page *Aij* and continuing on the verso of that page is the dedication to Maximilian by Martinus Ilacomilus (Waldseemüller). In that dedication, a reference to the accompanying maps is included: "Therefore, studying, to the best of my ability and with the aid of several persons, the books of Ptolemy from a Greek copy, and adding the relations of the four voyages of Amerigo Vespucci, I have prepared for the general use of scholars a map of the whole world—like an Introduction, so to speak both in the solid and projected on a plane."

A reasonable explanation for the two variant printings has been offered. The examples bearing the dedication of the Gymnasium Vosagense to Emperor Maximilian represented the publications to be sold commercially. The editions with Ringmann's address and Waldseemüller's dedications were authors' copies to be disseminated by the authors as part of their remuneration or as personal gifts. Of the extant copies, about one-third bear Ringmann's and Waldseemüller's names and most of them are known to have been owned by friends of the two men.[7]

An example of the title page of one of the editions, which were printed on April 25, 1507, is depicted in figure 4A. The translation is shown in figure 4B. (See pp. 96–97.)

Within the 104 small pages of *Cosmographiae Introductio*, several important assertions appear for the first time. The acceptance of Amerigo Vespucci as the discoverer of a new continent in the Western Hemisphere is the most pertinent. In addition, the title page announces the production of two maps—a gore map to be affixed to a sphere in order to create a globe and a plane (flat) map that would earn the designation of America's Baptismal Document.

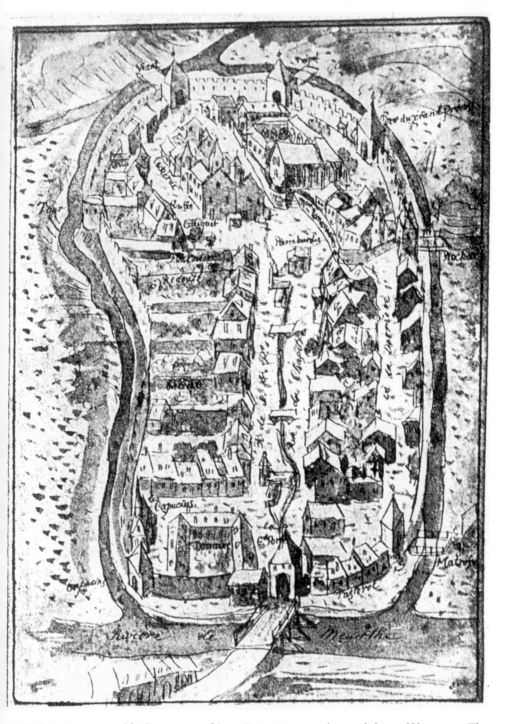

PLATE 1. Reconstituted bird's-eye view of Saint Dié as it appeared around the twelfth century. The fortified surrounding wall that was torn down by order of Cardinal Richelieu in 1633 was rebuilt. To the northeast is the quarter of the church and the residences of the canons. It is in this group of habitations to the left of the church that the "House of America" developed. Courtesy of Collection Médiathèque Victor Hugo, Saint Dié-des-Vosges.

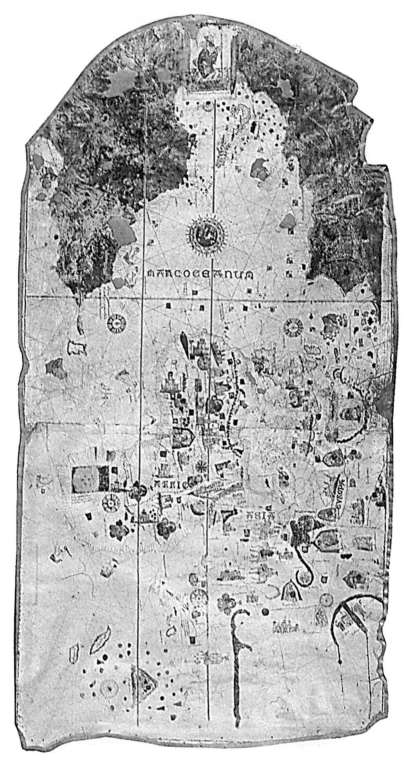

PLATE 2. Juan de la Cosa world chart, 1500, illuminated manuscript on parchment, 37.5 x 72 in. (960 x 1,830 mm). Courtesy of Museo Naval, Madrid.

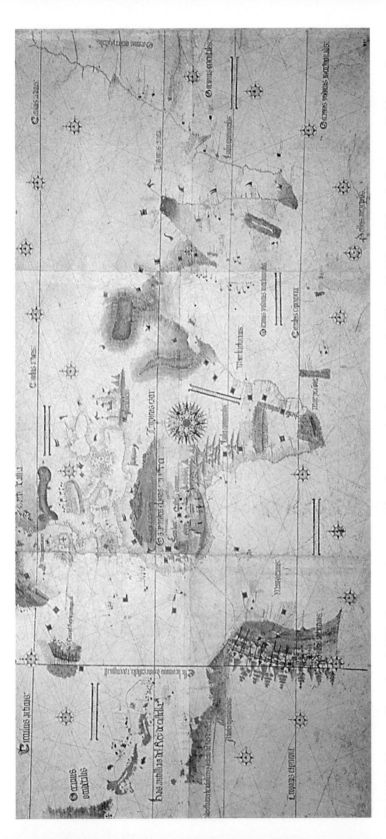

PLATE 3. "Cantino Planisphere," 1502, illuminated manuscript on three joined vellum sheets, 40 x 86 in. (11,202 x 2,180 mm). Courtesy of Biblioteca Estense, Modena, Italy.

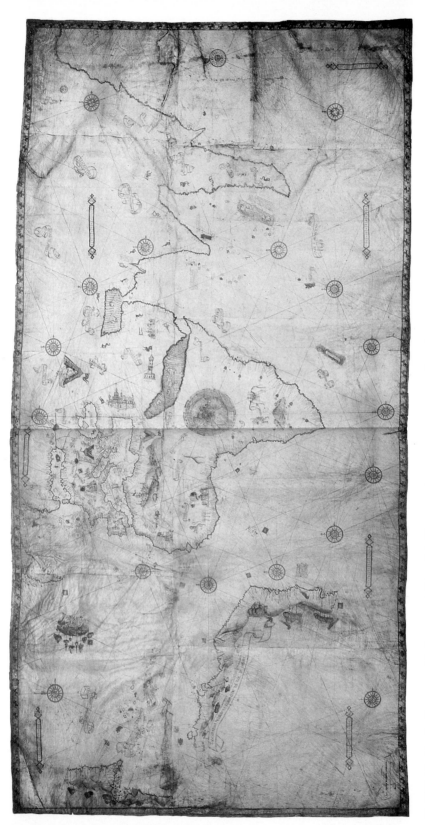

PLATE 4A. Nicolo Caveri world chart, ca. 1504–1505. Illuminated manuscript on ten vellum sheets, 46 x 90 in. (1,150 x 2,250 mm). Courtesy of Bibliothèque Nationale, Paris.

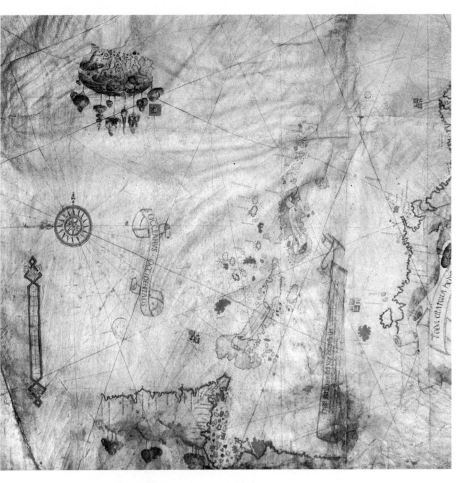

PLATE 4B. Detail of the map in plate 4A showing the Western Hemisphere.

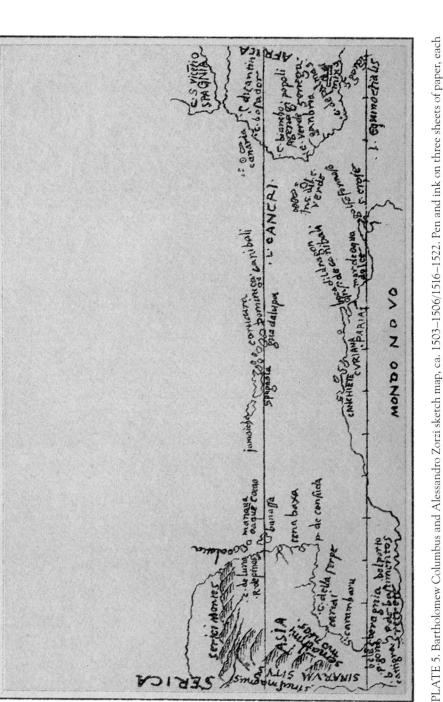

PLATE 5. Bartholomew Columbus and Alessandro Zorzi sketch map, ca. 1503–1506/1516–1522. Pen and ink on three sheets of paper, each 4 x 6.5 in. (100 x 165 mm). This is the sheet showing Christopher Columbus's discoveries in the Western Hemisphere. Courtesy of Biblioteca Nazionale Centrale, Florence.

PLATE 6. Giovanni Matteo Contarini-Francesco Rosselli world map, 1506, Venice or Florence. Copperplate engraving on paper, 17 x 25 in. (420 x 630 mm). Courtesy of British Library, London.

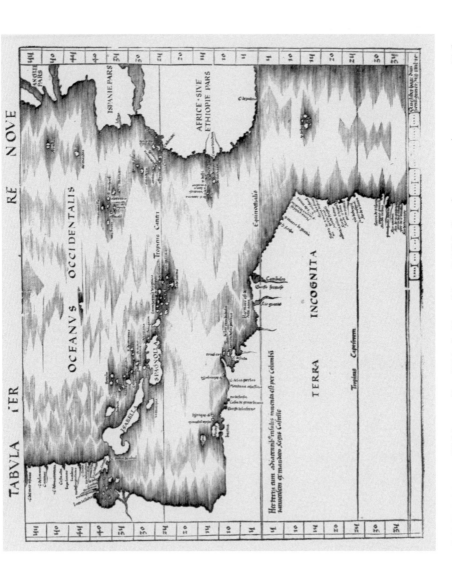

PLATE 7. Martin Waldseemüller, "Tabula Terre Nove," 1513, Strasbourg, woodcut on paper, 14.5 x 17.5 in (370 x 445 mm). From *Geographiae.*

The narrative of *Cosmographiae Introductio* specifically explains the reason for naming America and presents that name in print for the first time. The name appears twice, once in the text and once in the margin, apparently for emphasis, on the same page (fig. 5). The inclusion in the book of a complete Latin translation of Vespucci's four voyages is testimony to the importance assigned to that document by Martin Waldseemüller and his associates at the Gymnasium Vosagense. The name of Christopher Columbus is conspicuous by its absence. Since the narrative states that the purpose of the little book is to write a description of the two maps, *Cosmographiae Introductio* could well be regarded as providing the introductory notes or explication of the elegant document that graphically placed the name "America" on continental land in the Western Hemisphere.

The text begins with a presentation of the "Order of Treatment," in which the author indicates that he will include a discussion of the principles of geometry and astronomy to which will be added the four voyages of Vespucci "in order to form a basis for the description of the cosmography, both in the solid and on the plane." Thus, the narrative announced the production of two maps.

The first portion of the book is characterized by a running title at the top of the pages—"*Cosmographiae Rudimenta*"—and begins with a brief outline of the material that will be covered in nine chapters, then goes on to indicate that these chapters will be followed by a quadrant, the four voyages of Amerigo Vespucci, and a solid and plane map.

Chapter I, "Of the Principles of Geometry Necessary to the Understanding of the Sphere," presents a glossary of terms, which include definitions of circle, plane, circumference, center of a circle, semicircle, diameter of a circle, straight line, angle, right angle, obtuse angle, acute angle, solid, a degree, a minute (a sixtieth part of a degree), a second (a sixtieth part of a minute), and a third (a sixtieth part of a second).

Chapter II, "Sphere, Axis, Poles, Etc., Accurately Defined," expands on its title. The introductory paragraph refers to the importance of Vespucci's contribution: "Before any one can obtain a knowl-

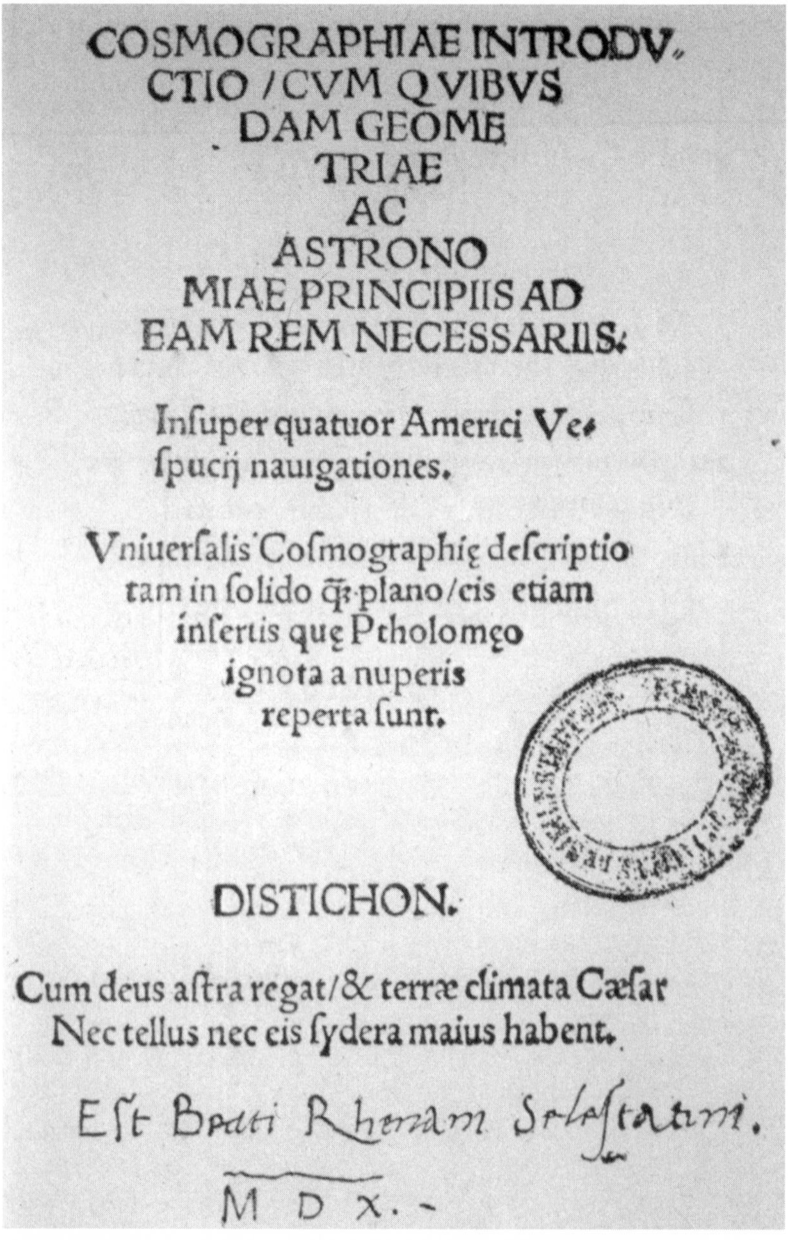

COSMOGRAPHIAE INTRODV.
CTIO / CVM QVIBVS
. DAM GEOME
TRIAE
AC
ASTRONO
MIAE PRINCIPIIS AD
EAM REM NECESSARIIS.

Infuper quatuor Americi Ve
fpucij nauigationes.

Vniuerfalis Cofmographiç defcriptio
tam in folido ꝗ-plano/eis etiam
infertis quę Ptholomęo
ignota a nuperis
reperta funt.

DISTICHON.

Cum deus aftra regat/& terræ climata Cæfar
Nec tellus nec eis fydera maius habent.

Eſt Beati Rhenam Selaſtatini.

M D X . -

Figure 4A. Title Page from *Cosmographiae Introductio*. Charles George. *The Cosmographiae Introductio of Martin Waldseemüller in Facsimile* (New York: United States Catholic Historical Society, 1907). Courtesy of Herbermann.

INTRODUCTION
TO COSMOGRAPHY

WITH CERTAIN NECESSARY PRINCIPLES
OF GEOMETRY AND ASTRONOMY

TO WHICH ARE ADDED

THE FOUR VOYAGES OF
AMERIGO VESPUCCI

A REPRESENTATION OF THE ENTIRE WORLD, BOTH IN
THE SOLID AND PROJECTED ON THE PLANE,
INCLUDING ALSO LANDS WHICH WERE UN-
KNOWN TO PTOLEMY, AND HAVE BEEN
RECENTLY DISCOVERED

DISTICH
Since God rules the stars and Cæsar the earth,
Nor earth nor stars have aught greater than these.

Figure 4B. Translation of the title page from *Cosmographiae Introductio.*

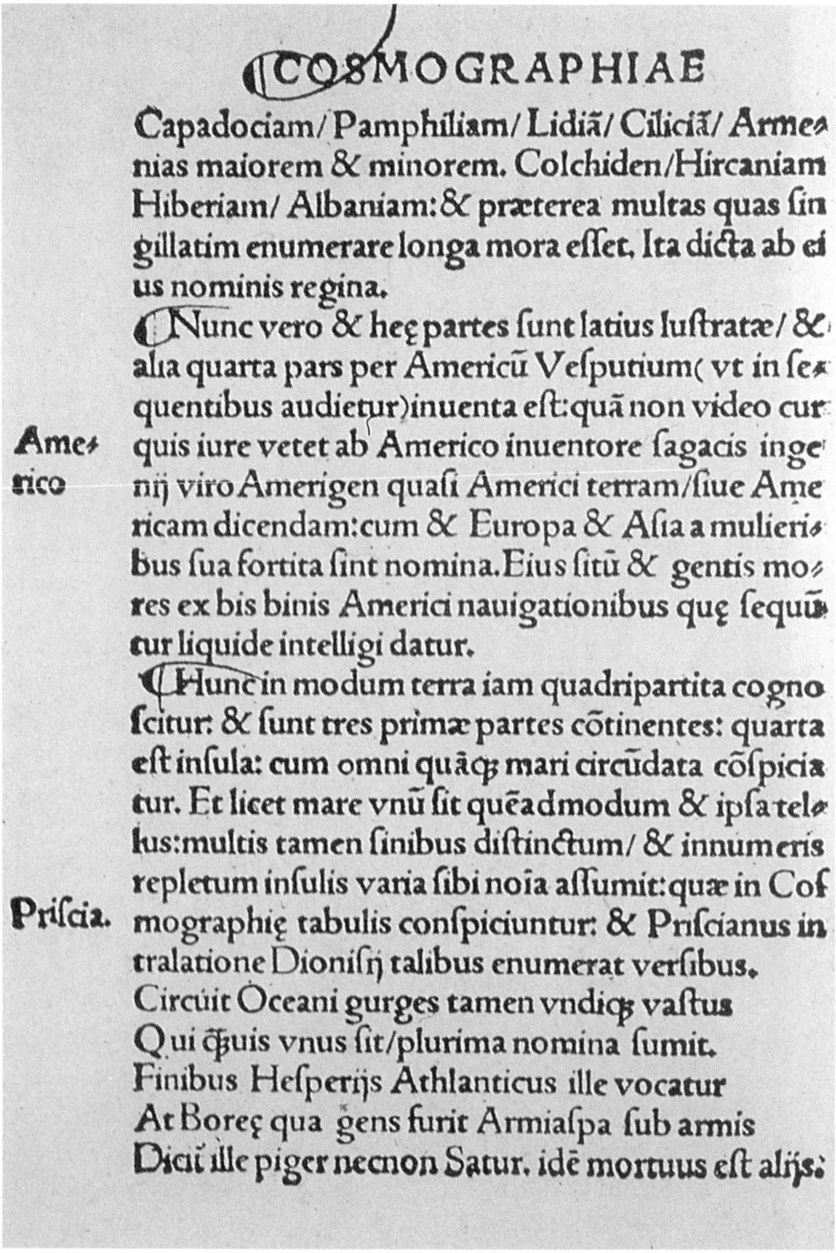

COSMOGRAPHIAE

Capadociam/ Pamphiliam/ Lidiã/ Ciliciã/ Arme◦
nias maiorem & minorem. Colchiden/Hircaniam
Hiberiam/ Albaniam:& præterea multas quas fin
gillatim enumerare longa mora effet. Ita dicta ab ei
us nominis regina.

  Nunc vero & heę partes funt latius luftratæ/ &
alia quarta pars per Americũ Vefputium(vt in fe◦
quentibus audietur)inuenta eft:quã non video cur
Ame◦ quis iure vetet ab Americo inuentore fagacis inge
rico nij viro Amerigen quafi Americi terram/fiue Ame
ricam dicendam:cum & Europa & Afia a mulieri◦
bus fua fortita fint nomina.Eius fitũ & gentis mo◦
res ex bis binis Americi nauigationibus quę fequũ
tur liquide intelligi datur.

  Hunc in modum terra iam quadripartita cogno
fcitur: & funt tres primæ partes cõtinentes: quarta
eft infula: cum omni quãcĝ mari circũdata côfpicia
tur. Et licet mare vnũ fit queadmodum & ipfa tel◦
lus:multis tamen finibus diftinctum/ & innumeris
repletum infulis varia fibi noia affumit:quæ in Cof
Prifcia. mographię tabulis confpiciuntur: & Prifcianus in
tralatione Dionifĳ talibus enumerat verfibus.
Circuit Oceani gurges tamen vndicĝ vaftus
Qui ĝĝuis vnus fit/plurima nomina fumit.
Finibus Hefperĳs Athlanticus ille vocatur
At Boreę qua gens furit Armiafpa fub armis
Dicū ille piger necnon Satur. idē mortuus eft alĳs:

Figure 5. *Cosmographiae Introductio*
page with "America(o)" in margin and text.

edge of cartography, it is necessary that he should have an understanding of the material sphere. After that he will more easily comprehend the description of the entire world which was first handed down by Ptolemy and others and afterward enlarged by later scholars, and on which further light has recently been thrown by Amerigo Vespucci."

Chapter III, "Of the Circles of the Heavens," defines the equator, the Arctic Circle, the Circle of Cancer, the Circle of Capricorn, and the Antarctic Circle. The Zodiac Circle is the celestial circle containing the twelve signs of the zodiac. The equinoctial circle, the solstitial, the meridians, and the horizon are also defined.

Chapter IV, "Of a Certain Theory of the Sphere according the System of Degrees," introduces the five principal circles: Arctic, Cancer, equator, Capricorn, and Antarctic. The distance between these circles is presented in degrees and minutes.

Chapter V, "Of the Five Celestial Zones and the Application of These and of the Degrees of the Heavens to the Earth," indicates that the Arctic and Antarctic zones are "frozen stiff with perpetual cold and uninhabited." The second and fourth zones, located respectively between the Arctic Circle and the Circle of Cancer, and the Circle of Capricorn and the Antarctic Circle, are temperate and habitable. The equatorial zone between the circles of Cancer and Capricorn is scarcely habitable on account of its heat. In this chapter, Vespucci is specifically credited with having recently discovered a very large part of the earth that had been unknown.

Chapter VI, "Of Parallels," defines parallels on the sphere as circles or lines that are equidistant from one another at every point. The definition precedes chapter VII, "Of Climates." "Climate" refers to a part of the earth between two equidistant parallels, "in which from the beginning to the end of the climate there is a difference of a half-hour in the longest day." North of the equator, seven climates have been given names from prominent cities, rivers, or mountains. Progressing north from the equator, these are Dia Meroes from Meroe (a city in Africa), Dia Sienes from Syene (a city in Egypt), Dia Alexan-

dria from the city of Alexandria, Dia Rhondon from the island of
Rhodes, Dia Rhomes from the city of Rome, Dia Borysthenes for the
river Borysthenes (now known as the Dnieper), and Dia Rhipheon for
the Rhiphæan Mountains in northern Europe. The most northern
eighth "climate," not mentioned by Ptolemy, is Dia Tyles from Thule.

South of the equator, the six climates that had been explored were
given the names of their northern counterparts: Antidia Meroes,
Antidia Alexandria, Antidia Rhondon, Antidia Rhomes, and
Antidia Borysthenes (Antidia Sienes was not listed in the text). In
this chapter, Amerigo Vespucci's discoveries on the coast of South
America are mentioned. It is stated: "In the sixth 'Climate' south of
the equator, closest to the equator, are included the tip of Africa,
Zanzibar, Java, and the fourth part of the earth, which, Amerigo discovered it, we may call Amerige, the land of Amerigo, so to speak, or
America." This reference constitutes the *first* appearance of the word
"America" in print. As John Boyd Thacher wrote, "This then, is that
simple sentence composed by an unknown geographer and printed in
an obscure town in a remote corner of the earth, which christened a
new world and fixed on it forever its pleasant sounding name."[8]

Chapter VIII, "Of the Winds," defines twelve winds related to the
summer rising and setting of the sun, the equinoctial rising and setting
of the sun, and the winter rising and setting of the sun to the north and
south of the equator. Four of the winds are referred to as "principal."
The east wind (Subsolanus) is pure and healthful. The west wind
(Zephyrus) is a mixture of heat and moisture and melts the snow. The
south wind (Auster) often precedes storms, hurricanes, and showers.
The north wind (Aquil) is cold and freezes the waters. The chapter
ends with a statement that "a general map indicating the poles, the
axes, the circles, great as well as small, the east, the west, the five zones,
the degrees of longitude and latitude, both on the earth and in the
heavens, the parallels, the climates, the winds, etc." is to be inserted.

On the back of the illustration, which appears after page XXVIII
in the edition reproduced by the United States Catholic Society, is a
definitive paragraph related to the maps called for on the title page of

Cosmographiae Introductio. It begins by stating, "The purpose of this little book is to write a description of the world map, which we have designed, both as a globe and as a projection. The globe I have designed on a small scale, the map on a larger." The remainder of the paragraph includes a description of the pictorial elements used to iden-tify specific regions. After describing symbols for locations in the Ptole-maic world, the author includes "and finally on the fourth division of the earth, discovered by the kings of Castile and Portugal, we have placed the emblems of these sovereigns . . . ," a direct reference to the New World. There, in the narrative of *Cosmographiae Introductio*, Waldseemüller's world map of 1507 is specifically announced! Chapter IX, "Of Certain Elements," is of the greatest significance. The text of this chapter presents the name "America" in print and indicates the reason for assigning the name and the source that served as the basis for the name. The introductory paragraph immediately indicates that the work represents a departure from previously accepted geography. "There is about a fourth part of this small region in the world which was known to Ptolemy and is inhabited by living beings like ourselves. Hitherto it has been divided into three parts, Europe, Africa, and Asia." The geography of Europe is defined and the derivation of the name from Europa, the daughter of King Agenor, is discussed. Simi-larly the geography of Africa is presented, as is the fact that the conti-nent was so named "because it is free from the severity of cold." Asia is described, and its name is ascribed to a queen of that name.

These considerations of the Ptolemaic world are immediately fol-lowed by the paragraph in which America is named: "Now, these parts of the earth have been more extensively explored and a fourth part has been discovered by Amerigo Vespucci (as will be set forth in what follows). Inasmuch as both Europe and Asia received their names from women, I see no reason why any one should justly object to calling this part Amerige, i.e., the land of Amerigo, or America, after Amerigo, its discoverer, a man of great ability. Its positions and the customs of its inhabitants may be clearly understood from the four voyages of Amerigo, which are subjoined."

In the margin of the page that contains the paragraph, the word "Americo" (America) appears (see fig. 5). The paragraph is followed by a statement that indicates that the continental land in the New World is surrounded by a great body of water. "Thus the earth is now known to be divided into four parts. The first three parts are continents, while the fourth is an island, inasmuch as it is found to be surrounded on all sides by the ocean." As indicated in the previous chapter, the insular concept is enigmatic, since at the time of the writing, there is no known description of the Pacific Ocean having been viewed previously by a European explorer from land, and there is no recorded voyage along the west coast of either North America or South America. The insular concept is generally considered to be speculative. The inclusion of the legend "the land beyond is unknown" on both northern and southern landmasses in the Western Hemisphere suggests that a link to Asia cannot be ruled out.[9]

Chapter IX continues with a consideration of the vast ocean that "surrounds the earth on every side" and the great seas of the Ptolemaic world. Larger islands are listed. A description of a method of determining the distance between one place and another follows. The chapter ends with a statement that there is a disparity between the representations as they are presented on the accompanying gore (globe) map and plane (flat) map. "We have therefore arranged matters so that in the plane projection we have followed Ptolemy as regards the new lands and some other things, while on the globe, which accompanies the plane, we have followed the description of Amerigo that we subjoin."

An appendix describes the method of constructing and using a quadrant to determine the elevation of the pole, the zenith, the center of the horizon, and the climates. An illustration is included. The entire section addressing "Cosmography" is concluded with a transitional statement: "Having now finished the chapters that we proposed to take up, we shall here include the distant voyages of Vespucci, setting forth the consequences of the several facts as they bear on our plan."

Of the 103 printed text pages making up the *Cosmographiae Intro-duction*, 63 consist of the translation from French into Latin of "The Four Voyages of Amerigo Vespucci." The section begins with the ele-giac and twenty-two-line poem by "Philesius Vosesigena" (Matthias Ringmann), which had previously served as the introduction to his *De Ora Antarctica*, bringing into focus the essence of the work, wherein lies the generally accepted basis for the naming of America: "Beyond Ethiopia and Bassa in the sea lies a land unknown to your maps, Ptolemy situated under the tropic of Capricorn and its com-panion Aquarius. To the right lies a land encircled by a vast ocean and inhabited by a race of naked men. This land was discovered by him whom fair Lusitania boasts of as her king, and who sent a fleet across the sea. But why say more? The position and the customs of the newly-discovered race are set forth in Amerigo's book. Read this, honest reader, with all sincerity and do not imitate the rhinoceros."

This introduction erroneously attributes the discovery of new land to Amerigo Vespucci but correctly refers to its encirclement by a large body of water. The basis of this geographic assessment is unknown. Yet another error is incorporated in the introduction. Vespucci is reported to have sailed under the flag of Portugal. In fact, the so-called first voyage—for which the wrong date is incorporated, thereby establishing priority of discovery—was conducted for Spain. The second and third voyages that are included in the translated text were sponsored by Portugal.

Ringmann's introduction is followed by a declaration that the translation was made from French into Latin. It is attributed to Jean Basin de Sendacour, who offers a ten-line poem followed by a two-line poem, informing the reader to anticipate a tale of the discovery of new lands during voyages conducted by Vespucci.

A complete English translation of the *Quatutor Americi Vespucij Navigationes* (Four Voyages of Amerigo Vespucci) is included as appendix II because the work constitutes the prime catalyst for the naming of America by Martin Waldseemüller and his associates at the Gymnasium Vosagense in Saint Dié.

The narrative of the *Lettera* indicates that Amerigo Vespucci sent the description of his travels directly to René II, the Duke of Lorraine. As professors Fischer and von Wieser point out in their assessment, it is unlikely that Vespucci would have sent an account from Portugal to the duke, with whom Vespucci had no personal relationship, and in French, the language that, according to the title page, was the basis of the translation. Reference by Vespucci to a time that he and the duke shared as students of Vespucci's uncle, Giorgio Antonio Vespucci, could not have applied to Duke René II. More likely, the original Italian letter that served as the basis of the printed document was addressed to Vespucci's fellow student Piero Soderini, who had become the gonfalonier (chief executive and minister of justice) of Florence. The reference to Dante as "our poet" is additional evidence that the original letter was written to an Italian rather than to the duke.[10]

According to the text, the first voyage, which was conducted by order of Ferdinand, king of Castile, began at Cadiz on May 20, 1497, and ended at the same port on October 25, 1499. Much of the narrative is a description of the natives and their lifestyles, constituting the first significant ethnographic dissertation on the natives of the Western Hemisphere. The country is identified as "Parias," thus giving Vespucci priority over Columbus as the first to set foot on a continent in the New World. But the date is certainly erroneous according to Alonso de Hojeda, who commanded the expedition, and also according to a manuscript letter of Vespucci, as will be noted in the next chapter.

In the narrative of the first voyage, landfall was stated to have been made at 16° north latitude. The description of the geography of the landfall is most compatible with Cape Gracias a Dios on the coast of Honduras. According to the text, the ships then proceeded northwest along the coast for 870 leagues or 3,100 English miles, which, in a direct line from Honduras, would reach to the longitude of Arizona or California—surely an exaggeration.[11]

According to the text in the printed narrative of *Lettera*, Vespucci left Cadiz on his second voyage in May "1489" (a print error for

"1499") and returned on September 8, 1500. It is generally agreed that the reported "first" and "second voyage," when combined, were based on the actual voyage that took place under the command of Alonso de Hojeda in 1499–1500.

The reported third voyage was conducted at the behest of King Manuel of Portugal. Vespucci set sail from Lisbon on May 10, 1501, with three ships. They made landfall in the Western Hemisphere at 5° south latitude. After visiting with the natives, the ships continued east and rounded the easternmost cape of South America and proceeded southwest to Cape St. Augustine (at 8° 21' south latitude). They then sailed south to about 50° south latitude before turning east. In 1502, after a voyage of sixteen months, the ships returned to Lisbon.

According to the text, the fourth voyage, conducted under the Portuguese flag, left Lisbon on May 10, 1503, and returned on June 28, 1504. The fleet encountered a large, uninhabited island, and the main vessel was wrecked on the offshore rocks. Vespucci continued west and reached land in the Western Hemisphere at 8° south latitude, where he left a contingent of two dozen men before returning to Lisbon. It was at Lisbon that the "Letter" describing the four voyages was supposedly composed and dispatched. But this was not possible, nor was the so-called fourth voyage, because *Vespucci was known to be in Seville at the time.*

The reader of *Cosmographiae Introductio* would end the experience with the viewing of a colophon bearing a cross with two horizontal bars and a vertical staff surrounded by the letters "S. D." for Saint Dié. The vertical staff extends into a circle, divided into three segments. The upper half is divided equally into two parts, one bearing the initials "G. L." for Gaultier Lud and "N. L." for Nicolas Lud. The lower half contains only the initials "M. I." for Ilacomilus (Martin Wald-seemüller), which are of a larger dimension than the other two. The illustration is surrounded by three statements: (1) This tome was printed and hereafter oft will others print, if Christ our helper be; (2) The town, St. Deodatus, named for thee and in the Vosgian Moun-tains reared aloft; and (3) Finished vij kl' Maij (April 25) 1507.

The reader, upon completing the work, would reflect on the newly acquired, unanticipated revelation of the contributions of Amerigo Vespucci, who by sailing westward from Europe discovered a continental mass surrounded by water. The name "America" was assigned to that continent. The narrative announced that accompanying maps would present a graphic representation of a newly appreciated revised geography. The reader was doubtless left with a sense of awe and anticipation.

CHAPTER 4
CREDIT AND CREDIBILITY

But in science credit goes to the man who convinces the world, not to the man to whom the idea first occurs.

Sir Francis Darwin

T he narrative embodied in *Cosmographiae Introductio*, by incorporating the "Four Voyages of Amerigo Vespucci" in its entirety and the accompanying two maps—one, the globe gores, and the other, Martin Waldseemüller's world map of 1507—effectively assigned Vespucci with priority as the discoverer of a New World in the Western Hemisphere. It further asserted that the continent, which he discovered, should honor him by bearing his name. The text and the two maps provided the culminating elements that resulted in what some have considered to be "The Greatest Misnomer on Planet Earth."[1] The basis of the designation "America," the attempts at erasure of that name, the establishment of its permanence, and the cynicism and vitriol that has been evoked over five centuries by the assignment of that name constitute a saga unto itself.

Our planet Earth derives its name from the Teutonic word for land. The larger continuous landmasses on Earth are referred to as continents. Three continents (Europe, Asia, and Africa) were included in the world that was described by the celebrated Greek astronomer Klaudios Ptolemaios of Alexander (CE 87–150), more commonly known as Ptolemy. Ptolemy introduced a method for constructing maps by suggesting projections, demonstrating how to break down a world map into component parts. He also incorporated the use of coordinates, latitude, and longitude to locate principal points on a map. Ptolemy's works disappeared from Europe during the Middle Ages but were preserved by Arabian scholars.

In the fifteenth century, with the collapse of Byzantium, Greek manuscripts, including the precepts of Ptolemy, were brought to Italy and played a role in the Renaissance. The text was translated into Latin, and the first printed edition of Ptolemy's *Geographia* was issued without maps in 1475 in Vicenza. The first edition to include maps (twenty-six copperplates based on Ptolemaic principles) was produced by Donnus Nicolaus Germanus in Bologna in 1477. Subsequent editions that were published before the production of the Waldseemüller map were printed in Rome in 1478, in Florence in 1482, in Ulm, Germany, in 1482 and 1486, and in Rome in 1490 and 1507.

In the Ptolemaic world, according to the narrative of *Cosmographiae Introductio*, Europe took its name from Europa, daughter of a Phoenician king. Mythology relates that Zeus, the king of the gods, was so impressed with Europa's beauty that he appeared as a bull and abducted her to Crete, where she bore him three sons, one of whom was Minos. In *Cosmographiae Introductio*, Asia is said to have derived from the name of a queen, Asia, the wife of the Titan Lapetus and mother of Prometheus, the Titan who stole fire from the gods. But the origin of the name of that continent is disputed, because Asia was also the name assigned by Virgil to a nymph in his tales, and the Koran notes that Asia was the name of the wife of the pharaoh who is credited with raising Moses. Etymologists, however, now suggest that continental "Asia" actually derives from *asu*, meaning "land of light," while Europe derives its name from *ereb*, meaning "setting sun" or "land of darkness."[2]

The third continent in the Ptolemaic world, Africa, was so named because it was devoid ("a") of extreme cold ("frica"). The Arctic, which is not a continent, took its name from the Northern Hemisphere's constellations, Ursa Major and Ursa Minor, the Great Bear and Little Bear. "Arkos" is Greek for "bear." The continent of Antarctica was so named because of its location opposite Arctica. It was thought that its presence afforded stability to the globe by balancing the ancient Arctica. Australia derives its name from "australis," meaning *southern*, because of its location in the Southern Hemisphere.

The two remaining continents, both in the Western Hemisphere, bear a variant of the Christian name of one man, Amerigo Vespucci. Thus, only one man's name is attached to any continent, and that name appears twice, on North America and South America. It must be concluded that Amerigo Vespucci is unrivaled in having his name perpetuated on planet Earth.

The life, the explorations, and the writings of Amerigo Vespucci (see figs. 6A and 6B) are cloaked in mystery. The man whose name was placed on two continents has been exposed to both praise and vilification.

Because no uniform calendar had been adopted at the time, the date of Amerigo Vespucci's birth is somewhat confusing. He was born, according to the baptismal record of the cathedral in Florence, the third of five children, on March 9, 1453, in the district of Ognissanti in Florence. According to the calendar in use in that city-state at the time, this would equate to March 18, 1454, in our current calendar. It is ironic that Vespucci was possibly born the same year that movable type was introduced, since it was two books printed with movable type that would play a critical role in establishing Vespucci's reputation. At his christening, which took place at the Baptistry of the Duomo in Florence nine days later, he received the name of his grandfather, Amerigo, which, in turn, derived from "Amalaric," the first Gothic king to occupy the throne of Seville. The baptistry's famous doors, referred to by Michelangelo as the Gates of Paradise, were completed by Ghiberti only one year before Vespucci's baptism.[3]

Amerigo Vespucci's family was a member of the bourgeoisie; his

Figure 6A. Amerigo Vespucci (1454–1512) portrait.
Engraving ascribed to Bronzini.

Figure 6B.
Statue of Amerigo Vespucci
in Uffizi Gallery, Florence, Italy.

father, Nastagio, was a notary of the Money-Changers Guild, as his father had been and his eldest son would be. His mother was Lisa di Giovanni Mini. The name *Vespucci* derives from the Italian and Latin *vespa*, meaning "wasp." The family coat of arms consists of a red field crossed by a blue band on which there is a procession of golden wasps. Amerigo was educated by his uncle, Giorgio Antonio, a Dominican humanist. Amerigo's fellow student was Piero Soderini, who would be made lifetime gonfalonier, or chief administrator, of Florence, and who was the recipient of Amerigo's critical letter concerning his discoveries in the New World.

In 1472, Nastagio commissioned one of the great Renaissance artists, Domenico Ghirlandaio, to paint the Vespucci family as a fresco for the wall of a chapel that was part of the church of S. Salvadore d'Ognissanti. Ghirlandaio, to whom Michelangelo was apprenticed as a thirteen-year-old, is remembered for his many frescoes; the most notable is the *Lives of the Virgin Mary and St. John the Baptist* in the Tornabuoni Chapel in Santa Maria Novella, Florence. On Ghirlandaio's fresco of the Vespucci family (fig. 7), to the right of the Madonna, is the only life portrait of Amerigo, then an eighteen-year-old youth.[4]

Although the circle of his associates included Machiavelli, Vasari, and Botticelli, Amerigo Vespucci's name does not appear until 1480, when the tax register of Florence states that "Amerigo, son of Ser Nastagio, aged twenty-nine, is in France with Messer Guido Vespucci [his uncle], ambassador." Amerigo Vespucci spent two years in France as an attaché and secretary to his uncle, who attended the court of Louis XI. After he returned to Florence, Vespucci became associated with the branch of the Medici headed by Lorenzo di Pierfrancesco de' Medici. That segment of the Medici family was known as *popolano*, because of its populist inclination in contradistinction to the more oligarchic portion of the Medici family.

Amerigo Vespucci looked after the business affairs of Lorenzo's family, and, in 1489, he made a short trip to Seville in order to investigate the status of the Medici's mercantile interests. His associations

Figure 7. Vespucci family; Fresco "La Pietà e Misericordia" by Domenico Ghirlandaio in the church of San Salvadore of Ognissanti, Florence. Amerigo Vespucci is in the lower scene behind his uncle and teacher, Giorgio Antonio.

there focused within the Italian colony that had developed as a commercial entity in Seville. Among the members of the Italian colony, a critical figure was Gianetto Berardi, a Florentine, who later became a close associate of Vespucci.[5]

It is quite possible that Vespucci and Christopher Columbus met during Vespucci's visit, because Berardi was friendly with Columbus, and Vespucci and Columbus shared a common admiration for the Florentine physician and humanist Paolo dal Pozzo Toscanelli, who was the first to suggest that the Orient and Spice Islands could be reached by sailing west from Europe.

Amerigo Vespucci's early interest in geography is attested to by his purchase of a portolan chart that had been made by Gabriell de Valsqua in Majorca in 1439. Vespucci wrote on the back of the map that he purchased it for 130 gold ducats. That same map was referred to by George Sand in *Un hiver à Majorque*, in which she stated that she viewed the map in the library of the count of Montenegro on Majorca while she and Frédéric Chopin were living on that island.[6]

Coinciding with the victory of Ferdinand and Isabella of Spain over King Boabdil and the Moors at Granada, Amerigo Vespucci left Florence for Seville in 1492 to look after the affairs of the *popolano* and Lorenzo de' Medici. Vespucci first became an agent for, and ultimately a successor of, Gianetto Berardi, who advanced Christopher Columbus money before he left on his first voyage. Although the major expense was underwritten by the Spanish sovereigns, the commander of the expedition was also expected to participate and share the financial risk. Shortly after Columbus's glorious return from that voyage, little time was lost preparing for a second voyage, for which Berardi and Vespucci coordinated the finances. Berardi served as a liaison between Columbus and the Spanish monarchs, as evidenced by the fact that he received power of attorney for Columbus when the "Admiral of the Ocean Sea" set forth from Cadiz on September 25, 1493, with seventeen vessels and about twelve hundred crew members on his second voyage.[7]

Berardi died at the end of 1495, and Amerigo Vespucci was left to act as executor for the estate and to continue the firm's activities related to Columbus's endeavors. When Columbus returned to Spain without any indication that he uncovered the anticipated wealth, he was rapidly divested of his heroic status. The financial losses incurred by the venture forced Vespucci to end his mercantile career. Left devoid of personal funds, Amerigo Vespucci turned to the sea.

Much of Vespucci's life at sea, and particularly the activities associated with the naming of America, is defined by the scholarly evaluation of two printed documents and three transcribed manuscripts. The two printed documents, to which Vespucci's name was attached and which played a crucial role in the creation of Vespucci as a legendary character, tell one story. By contrast, the earliest of the three manuscript letters written by Vespucci, which were discovered centuries after the naming of America, presents an entirely different impression. But before considering these disparities as important elements in the sequence of the naming of America, the undisputed remainder of Vespucci's life can be detailed briefly.

Amerigo Vespucci's initiation into the realm of exploration occurred during an expedition under the command of Alonso de Hojeda, who had served as a lieutenant during Columbus's second voyage. According to the writings at the beginning of the seventeenth century by Antonio de Herrera y Tordesillas, historiographer to King Philip of Spain, Vespucci went as a participant who "knew cosmography and matters pertaining to the sea" and also as a "merchant," representing the commercial interests of the backers of the expedition. Vespucci directed the course of two of the four vessels in the fleet. The ships under Vespucci's command separated from Hojeda and followed a more southerly course along the coast of South America.[8]

It is possible that in 1499, members of Vespucci's crew were the first Europeans to view the coast of Brazil. This circumstance would place Vespucci at the Brazilian shore seven months before Vicente Yáñez Pinzón, who is generally credited with the discovery of Brazil. This would also assign priority to Vespucci—not Pinzón—for discovery of the Amazon River. Some of the places appearing on the

1500 Juan de la Cosa map and the 1504–1505 Caveri map might have been named during Vespucci's voyage on that expedition. Most historians, however, do not include Vespucci's name anywhere on the list of those who are credited with discoveries in the New World, and his name does not appear on most rosters of sixteenth-century explorers.[9] There is no record of any voyage made by Vespucci for the king of Portugal in the *Torre do Tombo*, the general archives of Portugal.

But, according to the printed text in *Mundus Novus* and *Lettera* and the second and third manuscript letters of Vespucci, he participated on an expedition, possibly as navigator and astronomer, that was commissioned by King Manuel I of Portugal in 1501. During the voyage to the New World that lasted from 1501 to 1502, Vespucci named Cape St. Augustine at 8° south latitude on the coast of Brazil, and it is written that he proceeded as far south as 50° south latitude along the coast of Patagonia. From that point, Vespucci returned to Seville where he continued his commercial and maritime activities. He might have made two more voyages to the Western Hemisphere for Spain in 1505 and 1507. Reference to the first voyage appears only in a letter indicating that Vespucci and Juan de la Cosa had carried out a joint mission. There is also reference to Vespucci and Juan de la Cosa carrying gold from the Indies in 1508, a project in which Vicente Yáñez Pinzón and Juan Díaz de Solis are also said to have participated.

On March 22, 1508, Vespucci was appointed the first pilot major in the service of the Casa de Contratación of Spain. In that capacity, Vespucci was charged with the responsibility of teaching pilots his astronomical method of determining longitude and also constructing and maintaining the accuracy of the nation's all-encompassing official chart, known as the *padron general*, based on the recently reported voyages of discovery.[10]

Vespucci's reputation as an astronomer and navigator was based on his definition of the stars in the Southern Hemisphere made during his voyages along the coast of South America. He also bore the designation of "Astronomer to the King of Spain." The time spent as pilot major was abbreviated by poor health, probably a consequence of malaria. Amerigo Vespucci died in Seville on February 22, 1512, at the age of fifty-eight.

If we accept the fact that the naming of America stemmed from the activities at the Gymnasium Vosagense in Saint Dié during the first decade of the sixteenth century, it is not difficult to determine the course of events. Nowhere is the power of print more evident. William Caxton, the first great English printer, stated the truism "The spoken voice perishes, the written word remains."

Two printed documents appeared at the beginning of the sixteenth century and planted the seed for Vespucci's reputation and notoriety. The narratives presented Vespucci as a man who revolutionized geography by interposing a new continent between Europe and Asia. This refuted the concept of a single ocean that encompassed the landmass of Europe, Africa, and Asia. His name first appeared in print in a Latin translation of a letter (purported to have been originally written in Italian) that Vespucci had sent to his patron, Lorenzo di Pierfrancesco de' Medici. The letter, printed in a small book consisting of merely four leaves of paper, was titled *Mundus Novus* and was most likely first published in early 1503 in Florence (see appendix I). The time of publication is derived from the time of death of the addressee, which took place May 20, 1503. The book's popularity is attested to by the fourteen editions printed in Latin, among them a printing in Strasbourg prepared by Matthias Ringmann. Early editions were produced in Venice, Paris, and Antwerp, followed by many versions in other languages. The Augsburg Latin edition of 1504 was the first to bear a date. This was followed by printings in Cologne, Rostock, and a 1505 German translation in Nuremberg and a Flemish translation in Antwerp. The Latin editions were directed at an educated international audience, while the translations into national languages were for general readership.

The popularity and dissemination of *Mundus Novus*, a work that Vespucci was purported to have written, rivaled that of the famous letter from Christopher Columbus to Santángel. Columbus's letter, first published in 1493 with the title *De insulis in mari Indico nuper inventis*, constituted the first report of the islands in the Caribbean Sea. The readership of *Mundus Novus* was presented with the work at a time when Columbus's reputation had lessened. More important, the book contained the announcement of a "New World," a fourth continent. The terminology of a "New World" probably referred to the world unknown to Ptolemaic geographers. Although it appeared as "Mondo Novo" on Bartholomew Columbus's 1503 manuscript map depicting Christopher Columbus's fourth voyage, *Mundus Novus* was the first printed document to spread the name. The declaration of the existence of a New World was truly exciting, and the excitement was certainly amplified by the captivating description of the discovered land. The idyllic nature of the geography, coupled with the intriguing ethnographic descriptions spiced with inclusions of sexual activities of the natives, added to the appeal. What could have been more intriguing to the European readership than the sentence "When they were able to copulate with Christians, they were driven by their excessive lust to corrupt and prostitute all their modesty"?[11]

The statement "Albericus Vesputius to Lorenzo di Pierfrancesco de' Medici" introduces the narrative of *Mundus Novus*, thereby ascribing the authorship to Vespucci. A translation of the full text is presented as appendix I in order to underscore and allow the basis for the argument that the work is a *forgery*, and could not have been written by Vespucci himself.

The author of this work starts the letter by referring to a past letter written to the addressee concerning a prior voyage to new regions conducted for the king of Portugal. This is contrary to the report of the four voyages Vespucci described in the next publication attributed to him, that is, the letter to Soderini, where it is asserted that Vespucci's earlier voyages were under the flag of Spain. This error, however, is corrected later in the text. On one page the letter states that the author is describing a "first voyage," whereas later in the text, "two other (previous) voyages made under the mandate of the King of Spain" are referred to. The narrative is replete with inconsistencies concerning the coastal geography of the new continent and the time spent in transit along that coast. It is a known fact that Vespucci was at Cape Verde on June 4, 1501, the date of his meeting with the ships of Cabral's expedition. If, as is stated, it took sixty-seven days to complete the westward journey, it would have been impossible for his vessels subsequently to have sailed the reported three hundred leagues along the east coast of the newly discovered continent in three days.[12]

But, regardless of its errors and inconsistencies, the importance of *Mundus Novus* cannot be minimized. It defines the existence of a new "continent," "unknown to ancient authorities," south of the equator, across the Atlantic Ocean, and west of Europe. The narrative also offers the first ethnographic description of a continent in the Western Hemisphere. The land is characterized by a pleasant climate, fertile soil, abundant animals, and a population that exceeds in number the inhabitants of Europe, Asia, or Africa. Moreover, the text pointedly refers to a new hemisphere and describes, for the first time, some of the stars in the sky south of the equator in that hemisphere.

The impact of *Mundus Novus* is summarized by Stefan Zweig, who noted: "But the real fame and the world-historical importance of this tiny leaflet are caused neither by its content nor by the psychological tension which it creates among the people of the day. The actual event of this letter consists oddly enough not in the letter itself, but in its title—*Mundus Novus*—two words, four syllables, which revolutionize the conception of the cosmos as had nothing before."[13]

Mundus Novus refers to a voyage undertaken with the encouragement of King Manuel I of Portugal along the east coast of South America to 50° south latitude in search of a passage to the Orient from land that had been discovered by Cabral a year earlier. The narrative generally covers the material included in the second and third manuscript letters (to be discussed later). The same material is covered in the two Portuguese voyages, which were a part of the account of four voyages of discovery known as "Letter of Amerigo Vespucci concerning the isles newly discovered on his four voyages."

There is no question that the scholars at the Gymnasium Vosagense in Saint Dié had access to the printed material disseminated by *Mundus Novus*, because, as previously noted, one member of the group—Matthias Ringmann—edited an edition that was published in Strasbourg in 1505, just forty miles away. As is specifically indicated in the narrative of *Mundus Novus*, the author hoped to gather information from two previous voyages and a fourth anticipated voyage and to incorporate all of his travels in a book.

Mundus Novus introduced the name of Amerigo Vespucci to the group of scholars at Saint Dié, but it was a second book ascribed to Vespucci that cemented his reputation for the members of the Gymnasium Vosagense. Such a book appeared initially as a work in Italian titled *Lettera di Amerigo Vespucci delle isole nuovamente trovate in quattro suoi viaggi* (Letter of Amerigo Vespucci concerning the isles newly discovered on his four voyages) (see appendix II). The "Magnificent Lord" to whom the work is addressed has been identified as Piero di Tommaso Soderini, gonfalonier, or chief executive, of Florence, who had studied under the tutelage of Amerigo Vespucci's uncle, Giorgio Antonio Vespucci, at the same time as Amerigo.

A manuscript, written in Italian, was given to the bookseller Piero Paccini in Florence and was probably printed there in 1505 or 1506. There was minimal distribution of this printed edition, but a French translation was prepared and served as the basis for the Latin inclusion in *Cosmographiae Introductio*. The addressee of the original letter, "Magnificent Lord," was changed and appears in *Cosmographiae Introductio* as "To the most illustrious René, king of Jerusalem and of Sicily, Duke of Lorraine and Bar." The importance of this work, in the minds of the scholars at the Gymnasium Vosagense, is attested to by the inclusion of the full text in *Cosmographiae Introductio* (see appendix II).

In the *Lettera* (Letter of Amerigo Vespucci concerning the isles newly discovered on his four voyages), the two reported Spanish voyages generally correspond to the account in Vespucci's earliest manuscript letter. The most critical declaration of the four voyages, as they are presented in *Cosmographiae Introductio*, relates to the so-called first voyage, which, according to the text, was conducted under the Spanish flag from May 10, 1497, to October 15, 1498. It includes a landing in South America, thereby ascribing to Vespucci rather than Columbus priority for the discovery of a continent in the New World.

The first of the Portuguese voyages, which was referred to in the narrative of *Lettera*, is similar to the one described in *Mundus Novus* but adds the discovery of Cape St. Augustine and Bahia de Todos os Santos. The "Fourth Voyage" of 1503–1504 included in the text is incompatible with Vespucci's known presence in Spain at that time.[14]

Another printed work that contributed to Vespucci's priority as the discoverer of a continent in the New World appeared several months after the publication of *Cosmographiae Introductio*. Fracanzio da Montalboddo, a professor of rhetoric at the University of Vicenza, produced an Italian travel anthology titled *Paesi novamente retrovati. Et Novo Mondo da Alberico Vesputio Florentino intitulato* (Countries Recently Discovered and Called by Amerigo Vespucii the New World), which was printed in Vicenza on November 3, 1507. The work was reprinted at least seven times during the sixteenth century, translated into Latin by the monk Arcangelo Madrignano, and published in 1508 in Milan with the title *Itinerarium Portugallensium*. It was translated into French, published in Paris in 1515, and was finally included by Giovanni Battista Ramusio in the first volume of his widely read *Navigationi et Viaggi*, published in Venice in 1550.[15]

The anthology includes a description of several Portuguese expeditions, the first three voyages of Christopher Columbus, and a complete printing of the *Mundus Novus*, ascribed to Amerigo Vespucci. More important, the title of the anthology includes Vespucci's name but omits Columbus's. The error is compounded by the misleading title that could be interpreted as indicating that the "Countries recently discovered" were also discovered by Vespucci, rather than simply "called by Amerigo Vespucci the New World." This thereby reinforced the claim of priority of discovery for Vespucci.

Before tracing the evolution of the antagonism toward Vespucci, several manuscripts written during Vespucci's lifetime but discovered later merit consideration. In the middle of the eighteenth and the early nineteenth century, three manuscript letters were discovered in

Italy. In 1745, the Florentine abbot Angelo María Bandini published *Vita e lettere di Amerigo Vespucci gentiluomo fiorentino* (Life and Letter of Amerigo Vespucci, Florentine Gentleman), which included a letter from Vespucci to Lorenzo di Pierfrancesco de' Medici written in Seville and dated July 18, 1500, thus bringing to light a letter that had been forgotten for 245 years.[16]

The manuscript provides evidence for those who would absolve Vespucci as a perpetrator of a deliberate attempt to gain personal recognition as the discoverer of a continent in the New World. In the letter, Vespucci correctly indicates that his first voyage began on May 18, 1499, *a year after Columbus's landing on South America*. The text implies that Vespucci sailed as a pilot or navigator in charge of two of the four caravels in a fleet commissioned by the sovereigns of Spain under the command of Alonso de Hojeda. The document states that Vespucci proceeded south with two ships while Hojeda and Juan de la Cosa headed northwest with the other two vessels.[17]

According to the manuscript, on that voyage Vespucci supposedly discovered the mouth of the Amazon River and sailed upstream about forty-five miles. As he proceeded in a southerly direction, he encountered the Guiana stream and described for the first time that current, which runs from southeast to northwest to the east of South America. Vespucci continually studied the heavens and noted the stars of the Southern Cross. In the manuscript, the author describes the inhabitants he encountered on land in the "Torrid Zone" and, specifically, a race called "cannibals" who lived off human flesh. The author of the manuscript letter indicates that the vessels under his command sailed north to the Gulf of Paria, where the natives gave them pearls. He also describes exploring the island of Curaçao and notes many houses on stilts in the region, which reminded him of Venice. The name *Venezuela*, meaning "little Venice," appears on the Juan de la Cosa map. The naming of Venezuela is usually credited to Hojeda.[18]

The other two manuscript letters do not pertain to the issue of primacy of discovery in the New World. One, written by Vespucci from the island of Cape Verde to Lorenzo di Pierfrancesco de' Medici on

June 4, 1501, describes the activities along the coasts of Africa and Asia of Pedro Álvares Cabral's fleet. Vespucci gathered the information from the crews of several of the ships, which Vespucci met up with at Cape Verde. The third letter, written to Lorenzo from Lisbon in 1502, is essentially a continuum of the second letter and focuses on a description of the inhabitants, plants, and animals he encountered on the continent in the southern Torrid Zone of the New World. The letter concludes with the statement that more detail will be found in a book that Vespucci was preparing, titled *Voyages*.

The three private manuscript letters surely were not available to the scholars of the Gymnasium Vosagense at Saint Dié when they concluded that Vespucci merited recognition as the discoverer of continental land in the Western Hemisphere—a New World.

The narrative of *Cosmographiae Introductio* set the name "America" in print for the first time both in a statement explaining the basis for the naming and in the margin of the page bearing that statement. But it was the two accompanying maps—one a simple gore for a globe and the other an elegant flat map, which the title page and text of *Cosmographiae Introductio* referred to—that more dramatically presented a radically changed geography of the world and ingrained the word "America" on the land. The title of the Waldseemüller world map of 1507, "VNIVER-SALIS COSMOGRAPHIA SECVNDVM PTHOLOMÆI TRADI-TIONEM ET AMERICI VESPVCCI ALIORVQVE LVSTRA-TIONES" (World Description According to Ptolemy and the Travels of Americus Vespucius and Others), emphasized Vespucci's contribution to the expansion of the knowledge of geography.

It is a fascinating and informative fact that Martin Waldseemüller, who was responsible for placing the name "America" on maps, never included "America" on any of his subsequent maps. Waldseemüller is also credited with a leadership role in the production of the 1513 edition of *Claudius Ptolemaeus Geographia*. R. A. Skelton, superintendent of the Map Room for the British Museum, concludes: "After surveying the scattered and sometimes contradictory evidence, we can see no serious reason to doubt that Martin Waldseemüller was responsible for this grand and important edition (as Eames styled it), in the sense that he planned it, designed the new maps (except for that of Switzerland), wrote some if not all of the editorial texts, transported it from St. Dié to Strassburg, even perhaps helped to see it through the press."[19]

The 1513 *Geographia* represents the first production of Waldseemüller subsequent to the 1507 *Cosmographiae Introductio*, and there is documentary evidence that he and Matthias Ringmann were working on a new edition of *Ptolemy* as early as 1507 or even 1505. In *Speculi Orbis . . . Declaratio*, which was printed by Johann Grüninger in Strasbourg before April 1507, Gaultier Lud, the director of the Gymnasium Vosagense in Saint Dié, writes:

> But we would not deny that, in place of the enlarged delineation of Europe here given, might properly be inserted the representation, which we have hastily prepared of the unknown land discovered some time since by the King of Portugal, but subsequently to the preparation of this *speculum orbis*. Of the shores a fuller and more accurate representation will be found in the Ptolemy which— Christ willing—we shall soon publish at our expense, revised with many additions by Martin Ilacomylus [Waldseemüller], the most learned man in these matters.[20]

Another reference to a projected edition of *Ptolemy* appears in a letter dated April 5, 1507, signed by Waldseemüller, to Johann Amerbach, the humanist printer of Basel, stating: "I think you are aware that I am about to print in the town of St. Dié the *Cosmographia* of Ptolemy after revising it and adding some modern maps to it."[21]

The 1508 edition of *Ptolemy* incorporates a world map by Johann Ruysch, which includes a depiction of Columbus's discoveries and Cabot's North American discovery, shown connected to Asia, and Sylvanus's 1511 Venice edition, which presents Columbus's discoveries and two insular masses to the north discovered by the Corte-Reals. But the 1513 edition of *Ptolemy* is the first truly modern Renaissance edition that distinguishes between ancient and modern geography. The 1513 edition is divided into two parts, with the first part including the twenty-seven maps as usually shown in previous editions of *Ptolemy's Cosmographia* drawn on a trapezoid projection, and a second part containing twenty modern maps, which, with the exception of the map of northern Europe, are on a quadratic plane projection and are graduated in latitude but not longitude.[22]

Two of Waldseemüller's maps in the second part pertain to the Western Hemisphere. "Orbis Typus Universalis Iuxta Hydrographorum Traditionem" (see plate 21) is a world map that includes only the islands of "isabella" (Cuba) and "spagnolla" in the Caribbean Sea. No part of North America is shown; the South American continent is not named and contains only five coastal place names. The second map pertaining solely to the Western Hemisphere, "Tabula Terre Nove" (see plate 7), depicts many Caribbean islands, the southeastern portion of North America—including the suggestion of a Gulf of Mexico—and a Florida peninsula. Most pertinent, the South American continent no longer bears the name "America"; the name has been replaced with "Terra Incognita" (unknown land). A legend on the continent declares: "Hec terrra cum adiacentib[9] insulis inuenta est per Columb? ianuensem ex mandato Regis Castello" (This land with its adjacent islands was discovered by Columbus, sent by authority of the King of Castile). This suggests that Waldseemüller was correcting a mistake that he had made in 1507. There is no mention of Vespucci or the word "America" in the entire 1513 atlas.

The absence of reference to Amerigo Vespucci and the name "America" persists on another of Martin Waldseemüller's great maps. The twelve-sheet woodcut "Carta Marina Navigatoria Portugallen," printed in Strasbourg in 1516, contains the names "Terra Nova" and "Brasilia sive Terra Papagalli" (Land of Parrots) (see plate 20) on the northern and eastern portions of South America, respectively. A panel in the lower left of the map credits both Columbus and Vespucci with discoveries. In the Caribbean, no name appears on Cuba, and North America is designated as "Terra de Cuba-Asie Partis" in concert with Columbus's assertion that Cuba was merely an extension of Asia.

It is possible that Waldseemüller came to appreciate that Vespucci did not merit recognition as the discoverer of the mainland of South America and therefore erased "America" because of his reading of *Paesi novamenti retrovati*—published in 1507. The translation of that book into Latin, *Itinerarium Portugallensium*, the following year, or the High German translation, *Newe vnbekanthe landte* (New Unknown Lands), may have been the actual stimulus for the change. Each of these works states that the land was discovered first by Christopher Columbus, second by Petrus Aliares, and third by Albericus Vespucci.

But as is true in modern journalism, retractions are generally relegated to small print and are often neglected. In the case of the naming of America, this is obviously apparent. In 1515, Johann Schöner's globe, segments of the gores of which had been employed as folds in the large bound folio discovered by Professor Fischer in Wolfegg Castle in 1901 (see chapter 6), has the name "America" inscribed across the southern continent in the Western Hemisphere. The gores and an example of the globe, which is in the Grand Ducal Library at Weimar, Germany, are assigned the date of 1515. This is because it is believed that Schöner wrote the companion tract, *Luculentissima quaed terrae totius descriptio*, which is known to have been printed in 1515.[23]

The 1515 Schöner globe depicts the New World with two continental lands separated by a sea located between the Tropic of Cancer and the equator. The southern tip of South America forms the northern shore of a water passage between "Oceanus Occidentalis" and "Oceanus orientalis," which is shown seven years before Magellan's circumnavigation. This suggests an earlier Portuguese voyage through what became known as the Straits of Magellan. The separation of continental land in the New World from Asia is asserted by Schöner in his *Luculentissima*, in which he writes: "In this way it is ascertained that the earth is divided into four parts, and the first three parts are continents, that is mainlands; but the fourth [part] is an island because we see it surrounded on all sides by the sea." On the globe, the land in the northern section of the Western Hemisphere bears in large letters "PARIAS," which Schöner refers to in his accompanying narrative: "The island of Parias, which is not a part or portion of the said country, but a large special portion of the earth in that fourth part of the world." The appearance of the word "America" is also noted on a 1520 globe by Schöner.

The so-called Paris Green Globe, also assigned the date of 1515 and located in the Bibliothèque Nationale, Paris, presents the name "America" in four places and, for the first time, on the North American continent. Globe gores produced by Louis Boulengier in Lyons, France, between 1514 and 1517 name the South American landmass "America Noviter Reperta."

"America" reappeared in South America on a plane (flat) map in 1520 (see plate 8). That year, in Vienna, Peter Apian[us] published an obvious plagiarism of Waldseemüller's 1507 world map with "America" boldly imprinted on South America. On that map, 1497 is given as the date of Vespucci's first voyage. In 1522, Laurent Fries produced an edition of Ptolemy's *Geographia* that was printed at Strasbourg. This atlas presents reduced copies of the maps that Waldseemüller had drawn for the 1513 edition. Notably, on "Orbis Typus Universalis Iuxta Hydrographorum Traditionem Exactisime Depicta," in distinction to the original map, the name "America" is reintroduced on the southern continent in the Western Hemisphere. The name also appears on a circa 1527 woodcut hemispheric map by François Lemoyne (Franciscus Monacus), and on the portolan charts (manuscript sailing charts) of Gerolamo da Verrazzano (1539) and Vesconte de Maggiolo (1530). Peter Apian[us] also included the name on his heart-shaped world map of 1530, as did Johann Honter on his 1530 woodcut map. "America" is present on the 1531 double cordiform (heart-shaped) map of Orontius Finaeus, and is also present on the 1532 world map attributed to Sebastian Münster and Hans Holbein, the 1534 world map of Joachimus Vadianus, and the 1536 world map of Casper Volpell. The fate of the name "America" was probably sealed when it appeared on both North America and South America as "Americae pars Septentionalis & America pars Meridionalis" (North America & South America), on the 1538 double cordiform engraved world map of Gerhard Mercator (see plate 9)—one of the greatest cartographers of all time, noted for the projection bearing his name.

Although his name had been used to designate a continent, Amerigo Vespucci died without estate or glory in 1512, five years after his name first appeared on a map. At the same time, Christopher Columbus remained almost completely forgotten, seven years after his death in 1506. These two protagonists in the drama that focused on primacy of discovery never personally vied with each other. A letter written by Christopher Columbus to his son Diego in 1505, the year before Christopher Columbus died, indicates respect rather than animosity on the part of the Admiral of the Ocean Seas toward Vespucci. The original letter uncovered by Martín Fernández de Navarrete in the collection of the duke of Veragua, an indirect descendant of Columbus, states:

> My dear Son,
>   Diego Mendez departed here on Monday the third of this month.
>   Since his departure I have talked with Amerigo Vespucci, the bearer of this letter, who is going to court on some business connected with navigation. He has at all times shown a desire to serve me, and he is an honorable man. As with many others, fortune has not treated him kindly, and his labors have not been as rewarding as he deserves. He is going with a sincere desire to procure a favorable turn of affairs for me, if it is in his power . . .[24]

Also, it is known that Ferdinand, the natural son of Christopher Columbus, procured a copy of *Lettera* describing the four voyages of Vespucci. There is no suggestion that Ferdinand regarded the narrative as a meaningful attempt on the part of Vespucci to aggrandize Christopher Columbus's priority of discovery.[25]

About a half century after Vespucci died, the vitriol directed at him began. He was accused of being a usurper of Columbus's priority regarding the discovery of America by a most distinguished historian,

Bartolomé de Las Casas, the author of *Historia General de las Indias*. The manuscript of that work, which is preserved in the Biblioteca National in Madrid, was influential in its time, but was not published until 1875. Las Casas, whose father participated in Columbus's second voyage, was born in Seville in 1474. After his arrival in Hispaniola in 1502, Bartolomé de Las Casas became the first ordained priest in the New World and the first to be designated a bishop. He began writing the *History of the Indies* in 1527 and continued writing on his return to Spain until his death in 1566. In that work, Las Casas asserts:

> Now I can state with certainty that Vespucci was in bad faith and deliberately sought to steal the admiral's [Columbus's] glory . . . which proves that Vespucci was mistaken when he said he sailed in 1497. He sailed in 1499, and it is clear that Columbus discovered the continent. Which should have been called Columbia and not as it is unjustly called, America. (a footnote) It surprises me that the admiral's son, Hernando, who is such a wise man, did not notice how Americo Vespucci usurped the glory of his father, especially since he had documentary proof of it, as I know he has. . . . This then was the long premeditated plan of Americo Vespucci to have the world acknowledge him as discoverer of the largest part of the Indies.[26]

Las Casas summarized his case for assigning the priority of discovery to Columbus:

> Others besides these two (Pinzón and de Solís), say it is all one coast from Paria, though provinces have different names and there are also different languages. This, then, was declared by witnesses who had been there and knew it well by having used their own eyes, and now it would be needless to seek further for witnesses than in the grocers' shops in Seville. Thus it cannot be denied to the Admiral, except with great injustice, that he was the first discoverer of those Indies, so he was also of the whole of our mainland, and to him is due the credit, by discovering the province of Paria, which is part of all that land. For it was he that put the thread into the hands of the rest, by which they found the clew to more distant parts. Consequently, his rights ought

most justly to be complied with and respected throughout that land,
even if the region was still more extensive, just as they should be
respected in Española and other islands. For it was not necessary for
him to go to every part, any more than it is necessary in taking possession of an estate, as the jurists hold.[27]

Las Casas provided proof that Vespucci sailed under the command
of Alonso de Hojeda as "a merchant or an experienced seaman" in
1499, a year *after* Columbus had landed on the Paria Peninsula of
Venezuela. The 1601 publication *Historia de las Indias Occidentales* by
Antonio de Herrera y Tordesillas, historiographer to King Philip of
Spain, emphasized Columbus's role as discoverer of the New World's
southern continent. It concluded that Vespucci intentionally falsified
his report to seize personal glory. In his *Décadas*, Herrera writes: "With
great cunning Amerigo Vespucci transposes things that happened on
one voyage to another to conceal the fact that the admiral, Don
Christopher Columbus, discovered the mainland. . . . The invention
of Amerigo is clearly proved."[28] In 1627, Fray Pedro Simon expressed
his ire regarding the credit bestowed upon Vespucci and called for the
suppression of all geographical works and maps containing the name
"America."[29]

On the hundredth anniversary of the discovery of the New World,
Americae Retectio was published by Johann Stradanus. The title page
features parallel portraits of "Christophorus Columbus" and "Americus
Vespuccius," thereby assigning each equal credit. The three plates that
are included feature Columbus (fig. 8), Vespucci (fig. 9), and Magellan,
each presented aboard a ship surrounded by an allegoric setting.

Martín Fernández de Navarrete—one of the most authoritative
historians who studied the early voyages of discovery—published his
multivolume tome *Colección de los viages y descubrimentos* . . .
between 1825 and 1835. This work, which is one of the most complete collections of documents relating to the discoveries in the New
World, provided the background for many of the conclusions concerning the maritime activities of Amerigo Vespucci. It formed the

Figure 8. Christopher Columbus in *Americae Retectio*, Johannes Stradanus and Adrianus Collaert, Antwerp, ca. 1595.

Figure 9. Amerigo Vespucci in *Americae Retectio*, Johannes Stradanus and Adrianus Collaert, Antwerp, ca. 1595.

basis for the similar theories related to the priority of discovery of the New World, as espoused by Washington Irving and Alexander von Humboldt. Navarrete offered evidence that the first voyage of Vespucci could not have embarked in 1497 and that the fourth

voyage was a total falsehood. That evidence, however, has been criticized by Harrisse, with whom Arciniegas is in agreement.[30]

The vitriol directed at Amerigo Vespucci continued into the
nineteenth century, emanating from the pen of one of America's most
distinguished essayists, Ralph Waldo Emerson. In *English Traits*, he
compares the knavery and roguish characteristics of Saint George, the
patron saint of England, with the man whose name our nation bears:

> George of Cappadocia, born at Epiphania in Cilicia, was a low para-
> site who got a lucrative contract to supply the army with bacon. A
> rogue and informer, he got rich and was forced to run from justice.
> He saved the money, embraced Arianism, collected a library, and
> got promoted by a faction to the Episcopal throne of Alexandria.
> When Julian came, A. D. 361, George was dragged to prison; the
> prison was burst open by the mob and George was lynched, as he
> deserved. And this precious knave became, in good time, Saint
> George of England, patron of chivalry. Emblem of victory and
> civility, and the pride of the best blood in the land.
>
> Strange, that the solid truth-speaking Briton should derive from
> an imposter. Strange that the New World should have no better
> luck, that broad America must wear the name of a thief, Amerigo
> Vespucci, a pickle dealer at Seville, who went out in 1499, a subal-
> tern with Hojeda, and whose highest naval rank was boatswain's
> mate in an expedition that never sailed, managed in this lying world
> to supplant Columbus and Baptize half the earth with his own dis-
> honest name. Thus, nobody can throw a stone. We are equally bad
> off as our founders; and the false pickle-dealer is an offset to the false
> bacon-seller.[31]

An evaluation of the historic stature of Vespucci is dependent on
a reconciliation of the printed *Lettera* and a manuscript letter—both
attributed to Vespucci. In the printed document, his first voyage is
stated to have taken place in 1497, placing him on the Paria Peninsula of the South American continent a year *earlier* than Christopher
Columbus. By contrast, the manuscript letter states that he sailed

from Spain in 1499, thereby resulting in a landing on the South American continent a year *after* Columbus.

In 1879, Manning Ferguson Force read a paper before the Congrès international des américanistes in which he asserted that the printed *Lettera* incorporating the reports of four voyages made by Amerigo Vespucci was a forgery. He pointed out that, if there were any credibility to the claim that Vespucci landed on South American soil before Columbus, it would have been used by the crown in its defense against a suit brought by the heirs of Columbus, during which the government attempted to minimize the extent of Columbus's discoveries. In addition, if the friends of Columbus and his son Ferdinand had regarded the reports printed in *Lettera* as bearing any importance or weight, they would have loudly expressed their resentment. No such expression is noted. In 1895, Harrisse wrote: "The four voyages of Americus Vespuccius across the Ocean remain the enigma of the early history of America."[32]

The arguments over Vespucci's role continue into the twenty-first century. Professor Alberto Magnaghi, one of the foremost authorities on the life of Vespucci, concluded that both *Mundus Novus* and *Lettera* are spurious. In assessing the disparities between the printed description of four voyages and Vespucci's manuscript letters, particularly the earliest one, Magnaghi first proposed that the three manuscript letters represented translations of documents that originally had been written by Vespucci. As such, they provided truthful information and correct dates, specifically related to the first voyage. Thus, the printed *Lettera*, like *Mundus Novus*, has been designated as a "forgery." The number of voyages reported by "Vespucci" in *Lettera* is erroneous and might have been included to match the four voyages of Columbus, creating the sense of maritime equivalence between Columbus and Vespucci. Further evidence that the printed *Lettera* was not the work of Amerigo Vespucci is the fact that Piero Soderini would have been an unlikely addressee for Amerigo Vespucci because Soderini was in the political camp in opposition to Amerigo's patron, Lorenzo di Pierfrancesco de' Medici; and also Giovanni Vespucci, the son of Amerigo's cousin Guidantonio, participated in a plot against Soderini.[33]

The dates, the duration of parts of the voyages, and the identification of the latitudes of various locations, as they appear in *Lettera*, are all inconsistent and create confusion. The errors of geography contained in both *Mundus Novus* and *Lettera* are at odds with the expertise of a man who would become the first pilot major of Spain. The specific inclusion in the printed document *Lettera* of the name "Parias"—where Columbus made his continental landing in 1498—raises a suspicion of the motive for inclusion. The original Italian version of *Lettera* is replete with errors of language and is written in a style inconsistent with Vespucci's background and linguistic capability. The first two of the printed voyages appearing in *Lettera* might represent a composite of the two parts of an actual voyage on which Vespucci participated with Alonso de Hojeda in 1499–1500. This is dated correctly as extending from May 18, 1499, for thirteen months in a copy of a manuscript letter written to Lorenzo di Pierfrancesco de' Medici by Amerigo Vespucci from Cadiz on July 18, 1500.

Also, philologist and professor George Tyler Northrup presents textual evidence that supports the conclusion that the printed description could not be authentic. It has been suggested that the printed documents attributed to Vespucci should be labeled "para-Vespuccian," and that they were pirated from actual letters written by Vespucci to Lorenzo di Pierfrancesco de' Medici, altered, and amplified to attract a readership.[34]

Extreme vitriol is apparent in the statement referring to Vespucci published in 1931 in *Descobridores do Brasil*: "This fatuous personage is nothing but a lying novelist, a navigator of the caliber of hosts of others, a cosmographer who repeated the ideas of others, a false discoverer who appropriated the glory of others. Despite this, he managed to impress generations of learned men who spent their days trying to interpret fantasies and make sense of his nonsense."[35]

In defense of Amerigo Vespucci, Daniel Boorstin, the Librarian of Congress, indicated in his best-selling *Discoverers* that there has been a great deal of celebration of Columbus, whereas Vespucci has not had his due. The basis of Boorstin's statement was his conclusion that Vespucci "deserves fame as an opener of the modern mind, as the first man to declare that the land Columbus had stumbled upon and considered for all his life a part of Asia actually was a new continent, the fourth part of the world."

Until twenty years ago, the only known statue of Amerigo Vespucci in America consisted of his face and figure molded into the edge of a large bronze door that focused on Christopher Columbus at the Library of Congress. The door was destroyed by fire in 1851 and never replaced. Currently, near the library checkout counter at Glendale Community College in the hills north of Los Angeles a bust of Amerigo Vespucci stands on a marble pedestal. This is the culmination of a drive generated by Dr. Putnam Kennedy, a retired radiologist, who persuaded the Kiwanis Club of Glendale to sponsor the project. In 1986, Mexican sculptor Armando Amayo produced the bust. Before the mold was broken, two additional castings were made. One statue was placed at the Colombian Academy of History in Bogotá, and the other was unveiled in a park in Rio de Janeiro. In 1987, a Bogotá company commissioned another large statue of Vespucci for that city.

If we sift through the arguments and compromises, some conclusions can be reached regarding the issues of priority and calumny. Christopher Columbus certainly was a great seaman—the first to set foot on the South American continent during his third voyage in August 1498. However, he persisted to his dying days with the belief that he had reached the East Indies and the Orient and never considered that he had discovered a continent that was interposed between the Atlantic Ocean and the Sea of Cathay (China). Amerigo Vespucci, who was not noted for his seamanship, was a knowledgeable navigator and astronomer. He reached South America in 1499, a year *after* Columbus, as a participant in an expedition led by Alonso de Hojeda. He is to be credited with the conviction that the newly discovered land was a New World, and that there was continental mass distinct from Asia and the East Indies.

It could be argued that Columbus does not deserve credit for Waldseemüller's revolutionary geographic presentation—of a new continent—interposed between two large seas. It was the printed documents attributed to Vespucci that brought that concept to light. Although Columbus found land by sailing westward from Europe and crossing the Atlantic, the Vespuccian narratives, whoever authored them, established the presence of a new continent and a Western Hemisphere.

Vespucci also made significant astronomic observations of the skies south of the equator. Accepting the veracity of his earliest manuscript letter discounts any part Vespucci may have played in an attempt to usurp Columbus's priority of discovery of South America and thereby to advance his own stature. The facts presented in the printed documents represent questionable authorship as well as accidental or intentional alterations by the publishers. As Washington Irving suggested: "It may have been the blunder of some editor, or the interpolation of some book-maker, . . . to gather together some disjointed materials, and fabricate a work to gratify the prevalent passion of the day."[36]

Given that "America" is a misnomer based on a printed error granting Amerigo Vespucci priority of discovery, it is intriguing to consider other names that could possibly have been applied to the continental land in the Western Hemisphere. The southern continent might have honored Christopher Columbus and could have been named Columbo, Colon, Columbia, or Columbus. If, however, the pattern to be followed drew from the past when the given names of women were used, a variant of Christopher, such as Cristofero, would be more appropriate. Alas, the only significant locales honoring the Admiral of the Ocean Sea are a relatively small South American country, the capital of South Carolina, and the district in which the capital of the United States of America is located. In the latter two instances, the names can be traced to a term, "Columbia," used by Philip Freneau in his poem "American Liberty," published in 1775. "Columbus" was first used in 1812 when it was applied to the newly designated capital for the state of Ohio.

Because neither Columbus nor Vespucci had any contact with North America, it might have been historically correct for Renaissance Europe to honor Giovanni Caboto or to use the anglicized *Cabot*. No major community in North America honors this discoverer in North America. But Cabot *did not set foot* on North American continental land. That priority is generally assigned to Ponce de León, who arrived at Florida in 1513. His name is absent from our continent, appearing only on the southern shore of Puerto Rico, where he served as governor. In 1521, Gordillo and de Quexos made an inconsequential landing at Winyah Bay, South Carolina. Their names appear on no maps. Giovanni da Verrazzano is credited as the first European to define the east coast of North America from northern Florida to Cape Breton. He is honored only by the Narrows that provide entrance to New York Bay, which he explored in 1524, and the bridge that spans the Narrows.

Yet other errors were made in assigning the Christian name of "Vespucci" to land in the New World. If we assume that the Latin equivalent of "Amerigo" was more appropriate, the correct version would have been "Albericus," not "Americus." And if it had to be "America" because it sounded better, an accent should have been placed on the second syllable: "América."

The word "American" was first applied to inhabitants of the land in 1578 in a description of the voyages of Martin Frobisher, who sailed in search of a northwest passage. According to Frobisher, "They are of the colour of a ripe olive, which is how it may come to passe, being borne in so cold a climate, I referre to the judgement of others, for they are naturally borne children of the same colour and complexion that all the *Americans* [my italics] are, which dwell under the Equinoctiall line."[37]

The German American Heritage Society of Greater Washington, DC, asserted in its December 2001 publication "America's Baptismal Certificate" that, because America was named by a German scholar and mapmaker, it is appropriate to consider the German derivation of that name. *Amerigo* is an Italian version of the Old German name

Haimirich that the Ostrogoths (East Goths) introduced to Italy in the fifth century. "[L]ater forms were *Emmerich, Emeric, Emery, Ameryk,* and *Amery.* The roots of the name are the Germanic *haimi, heim, home* and *rikja, reiks, ric, rich, reich,* meaning power or ruler. . . . Thus our continent was not only named by a German but also carries a name with a Germanic derivation."[38] Ogden Nash applied his distinctive wit to the issue of Amerigo Vespucci's role in the naming of America in his poem "Columbus":

> So Columbus said, show me the sunset and
>   somebody did and he set sail for it,
> And he discovered America and they put him in jail for it,
> And the fetters gave him welts,
> And they named America for somebody else,
> So the sad fate of Columbus ought to be pointed out to
>   every child and voter,
> Because it has a very important moral, which is Don't
>   be a discoverer, be a promoter.

There is yet another somewhat obtuse interpretation regarding the origin of the name "America" and its attachment to land in the New World. During the twentieth century, recurring suggestions with blatantly chauvinistic undertones indicate that America might bear a name that actually derives from a Bristol merchant, Richard Amerike.[39]

When John Cabot—sailing under the auspices of a trading license that was issued by King Henry VII of England—left Bristol on the *Matthew* with a crew of twelve on May 2, 1497, the charge was primarily to establish trade with Cathay. The voyage was to proceed along a northern latitude so that it didn't antagonize Spain, which had already evidenced interest in the region of the Caribbean Sea. At the time, Henry VII was negotiating a marriage between his son Arthur and Catherine of Aragon of the Spanish court.

The major investor in Cabot's voyage was Richard Amerike, a

senior member of the Fellowship of Merchants and the customs officer of Bristol. After an uncomplicated voyage across "Mare Oceanum" (Atlantic Ocean), Cabot and his crew went ashore in the region of Newfoundland or Cape Breton on June 24, 1497, and planted the Royal Ensign, claiming the land in the name of Henry VII. This discovery occurred more than a year before Christopher Columbus landed on the Paria Peninsula of Venezuela. Like Columbus, Cabot nurtured the opinion that he reached an island off the coast of Asia or, perhaps, the mainland itself. After Cabot returned to Bristol on August 6, 1497, he went directly to London to report it to the king, who named the territory "New founde land."[40]

Toward the end of the nineteenth century, in the Chapter House of Westminster Abbey, Edward Scott discovered documents that detailed the custom collections and expenditures for the city of Bristol for 1498 and 1499. When British historian Alfred Hudd reviewed the documents, he noted that the name "Richard Ameryk" appeared on one and "Richard ap Meryke" on the other. These names undoubtedly referred to the same individual because the name *Ameryk* derives from the Welsh, *ap Meryk*. Other documents provide evidence that Richard Amerike, in search of cod, played a key role in the voyages that sailed from Bristol in the 1470s and 1480s to the island of Brassyle, about four hundred miles west of Ireland.[41]

These facts led Hudd to conclude that Cabot might have named America for Richard Amerike. Hudd presented his thesis at a meeting of the Clifton Antiquarian Club in Bristol on May 21, 1908. After his death, the *Western Daily Press* of Bristol, on August 7, 1929, printed the conclusion of Mr. Hudd's address:

> There is no longer any doubt on the return of [Cabot's] second voyage John received for the second time the handsome [sic] pension conferred upon him by the King, from the hands of the Collectors of Customs of the Port of Bristol. One of these officials, the senior of the two, who probably was the person who actually handed over the money to the explorer, was named Richard Ameryk (also written Ap Meryke in one deed) who seems to have been a leading

citizen in Bristol at the time, and was Sheriff in 1503. Now it has been suggested by Mr. Scott and myself that the name given to the newly-found land by the discoverer was "Amerika," in honour of the official from whom he received his pension. [42]

The *Bristol Evening Post* of April 30, 1943, includes the lead article titled "AMERYK AND AMERICA." The text states:

Although we like to think that Cabot discovered America—and there seems historical evidence in support of the fact—most Americans seem to favour Columbus' claim.

I was not a little surprised therefore to hear Gen. H. R. Ingles, presenting the Stars and Stripes to the Lord Mayor, say: "Cabot was really the discoverer of what is now known as America."

Concerning that statement, I have to thank Mr. H. R. Simpson (M.A., Camb.) for an interesting contribution.

"I think you could assure Americans that they owe not only their name but their flag to Bristol," he writes. "The accepted derivation of the word America from Amerigo Vespucci is based on evidence so slight as to be also fantastic. The name America was given by Richard Ameryk, Lord of the Manor of Clifton, and sometime Sheriff and Receiver of the King's Customs at Bristol.

In any event, he financed Cabot, possibly with the help of his son-in-law, Broke, and it seems inevitable that he should have stipulated that any new continent disclosed should bear his name.

Cabot, an unreliable character, broke the bargain. He called his discovery Newfoundland—not an inspired name. Ameryk took the only revenge open to him. When as the Receiver of the King's Customs, he was ordered to pay Cabot the reward of £10, Ameryk refused to pay. Finally, when a very summary order came from the Exchequer, he did pay—and it is shortly after that that the name America begins to be used.

Any encyclopedia will give the derivation of the Stars and Stripes, but the fact remains that the arms of the Ameryk family were stars and stripes. A curious coincidence, to say the least of it."[43]

In 1951, Bernardini Sjoestedt in *Christophe Colomb* also attributed the origin of the name "America" to Richard Amerike, the financier of Cabot's voyage. This theory is difficult to accept because the name "America" does not appear in the documents written shortly after Cabot's exploration. Much of the material that has been referred to ascribing the name "America" to a Bristol merchant recently resurfaced in *Terra Incognita: The True Story of How America Got Its Name* by Rodney Broome, who lives and works in Seattle, Washington, but was born and raised in Bristol.

In the fifteenth and sixteenth centuries, Bristol was the second-largest city in England and a major seaport. The city was built on a large bend of the Avon River. Twice daily, tides in the Bay of Cardiff and the Bristol Channel rise and fall thirty feet. The city was granted its seal in 1350. In the tenth century, Bristol was known as Brygestowe, derived from "bryge," meaning *bridge*, and "stowe," meaning *meeting place*.

Let us consider all the facts that indicate that America bears the name of Amerigo Vespucci, and that the thesis that Richard Amerike is the honoree is purely conjectural and almost exclusively promulgated by men from Bristol. It would thus require a bridge with a long span to transcend the shelves of evidence that we've been examining. It must therefore be concluded that the name "America" was specifically assigned by Martin Waldseemüller to honor Amerigo Vespucci.[44]

CHAPTER 5

DISAPPEARANCE
AND DISCOVERY

Let us look at the map, for maps, like faces, are the signatures of history.

Will Durant

Before its discovery, the Waldseemüller world map of 1507 had been referred to as "the Holy Grail of American cartography." The symbolism of the Holy Grail itself was introduced in medieval times: it supposedly was the chalice from which Jesus Christ drank during the Last Supper, and it became the quest of many chivalrous expeditions. The word *grail* derives from an Old French word that referred to a flat plate. An interesting similarity between the Holy Grail and the long sought after cartographic jewel is that the map was specifically described as a "plane," or flat, map. The quest for the Holy Grail was a treasure hunt, however, with no clues and no certainty of the object's existence. By contrast, the quest for the holy grail of American cartography was undertaken with strong evidence that the map actually existed, and there was a pervasive anticipation that an example of the map eventually would be found.

As previously noted, in 1828 Washington Irving was the first to

trace the earliest appearance of the name "America" to the text of
Cosmographiae Introductio, and he was the first to point out that the
existence of the map was boldly announced on the title page of that
work. On April 25, 1507, when that small book came off the press of
the Gymnasium Vosagense at Saint Dié, the title page stated that an
integral part of the printed production was "A Representation of the
Entire World, both in the Solid and *Projected on the Plane*, Including
also Lands which were Unknown to Ptolemy, and have been Recently
Discovered." This announcement was reinforced by the cartographer
of the gymnasium, Martin Waldseemüller, in his personal preface to
the *Cosmographiae Introductio*, in which he addressed His Majesty
Maximilian Cæsar Augustus: "I have prepared for the general use of
scholars a map of the whole world—like an introduction, so to
speak—both in the solid and projected on a plane." Reference to the
existence of the map also appears in several passages in the narrative
of the book. In the section on cosmography, the "Order of Treatment"
states: "Thus we shall describe the cosmography, both in the solid and
projected on the plane."[1]

In chapter VII, "Of Climates," in the *Cosmographiae Introductio*,
the following sentence appears: "Our world map, for the better understanding of which this is written, will clearly show you the beginning,
the middle, and the rest of this first climate and also the rest, as well
as the hours of the longest day in every one of them." At the conclusion of chapter IX, "Of Certain Elements of Cosmography," reference
is made to the fact that the map consists of several sheets: "All that
has been said by way of introduction to the Cosmography will be sufficient, if we advise you that in designing the sheets of our world map
we have not followed Ptolemy in every respect, particularly as regards
the new lands, where on the marine charts we observe that the
equator is placed otherwise than Ptolemy represented it."

Also, within the *Cosmographiae Introductio* that came off the small
Saint Dié press in 1507, on the back of a drawing of a globe with its
principal lines of direction, a text states: "The purpose of this little
book is to write a description of the world map, which we have

designed, both as a globe and a projection. The globe I have designed on a small scale, the map on a larger." This is followed by a description of the emblems that appear on the map, each representing a specific realm. The detail suggests a large, possibly colored map consisting of multiple sheets. Thus, acquaintance with the narrative of the *Cosmographiae Introductio* that refers throughout the work to a map and, specifically, a large map of several sheets, led to the anticipation of finding a monumental map of considerable size.

The existence of such a map was substantiated in a letter dated August 12, 1507, from the well-known Benedictine historiographer Johann Trithemius to Veldicus Monapius. In that correspondence, the writer indicates that, a few days earlier at Worms, he bought cheaply a "handsome terrestrial globe of small size lately printed at Strassburg, and at the same time a large map of the world containing the islands and countries recently discovered by the Spaniard, Americus Vespucius, in the western sea, which map extends south almost to the 50th parallel . . ." Father Joseph Fischer, the man who discovered the map in 1901, wrote in his definitive description of his find: "From the results of our researches thus far, we do not in the least doubt, that Trithemius here speaks of the three parts of the joint-publication of Waldseemüller of the year 1507." Harrisse, in his classic compendium of the early mapping of America, produced before the Waldseemüller world map of 1507 was discovered, also expressed the belief that the map referred to by Trithemius was, indeed, the long sought after Waldseemüller large map.[2]

Henricus Glareanus (Loritus Loritz), a Swiss humanist, mathematician, geographer, and astronomer, who later taught poetry at the University of Freiburg, also attested to the existence of the Waldseemüller map. In 1892, Franz Ritter von Wieser—who would announce the discovery of the Waldseemüller map nine years later—found two small pen-and-ink colored drawings, about 19.5 × 26 centimeters (8 × 10 inches) by Glareanus inserted into a copy of *Cosmographiae Introductio*, which was located in the Universitätsbibliothek in Munich (see plates 10A and 10B). This extraordinary work was purchased by the library from a Mr. Rosenthal, sometime before 1890 for 300 reichsmarks, about $1,500 at the time or $12,000 in the current equivalent.

The two primitively sketched maps, dated 1510 by Glareanus, contain the name "America" on the southern continent in the Western Hemisphere. In a legend on one of the maps, Glareanus writes that the maker of the original map drew it so large that it would not fit in the codex (book), and, therefore, Glareanus drew his rendition of the habitable world in proportion to the larger original but in a smaller space. The map was stated to show "three parts of the world and the fourth American part recently discovered."[3]

In 1896, A. Elter found another copy of the two Glareanus maps inserted into a 1482 Ulm edition of Ptolemy in the University Library in Bonn. The notes made on one of those maps by the copyist indicate that he followed "the geography from the Deaodatans (Saint Dié) also known as Vosgiens." The map of the Western Hemisphere contains the name "Parias" on the northern continent, "Isabella" and "Spagnolla" in the Caribbean, and "Terra America" on the southern continent. Joseph Fischer and Franz Ritter von Wieser used the Glareanus manuscript maps as proof of identity of the Waldseemüller world map. These scholars cited as evidence agreement in projection and contour of the lands as depicted on the Glareanus maps and the Waldseemüller map, which they had recently discovered. Also, they pointed out that writings on the Glareanus manuscripts made direct and specific reference to the original Waldseemüller map as the source.[4]

Additional evidence that led to the anticipation of the existence of the Waldseemüller world map of 1507 appeared in Kraków in 1512. Included in the *Introductio in Ptholomei Cosmographiam* by Johannes Stobnicza, the Stobnicza map is a woodcut, measuring 27×38 centimeters (11×15 inches) (see plate 11), on which the New World is shown with a continuous coastline from 50° north latitude to 40° south latitude. Like the Waldseemüller map, Stobnicza's map depicts continental land in the New World interposed between two oceans and separate from Asia. Stobnicza published his map without mention of a source, but there is little question that both Eastern and Western hemispheres were taken directly from the two inset maps at the top of the Waldseemüller map. On the Stobnicza production, the island of "Spagnolla" (Hispaniola) is named and "Isabella" (Cuba) is shown as part of the North American mainland. The only other place named on the northern continent is "Cabo de bona ventura." The southern continent is named "Terra Incognita" (Unknown Land). The other seven place names on the southern continent also appear on the Waldseemüller map, but the word "America" is absent. The primitive nature of the Stobnicza map suggested a more sophisticated source. Before the discovery of the Waldseemüller map, Harrisse wrote: "The relationship between the Polish mappamundi and Waldseemüller's map [1513] is further shown by the inscriptions. . . . It is the evidence of a continuous coast in the map of Stobnicza which led us to imagine that this configuration may date further back than the construction of Waldseemüller's lost planisphere (projection of a sphere on a flat map) of 1507."[5]

Several globes and gores produced during the first quarter of the sixteenth century offered a revolutionary world geography, thought to have been based on the map called for by the *Cosmographiae Introductio*. Edward Stevenson, a distinguished authority of terrestrial and celestial globes, referred to a gore map in the library of the prince of Liechtenstein. The map, which now is a highlight of the James Ford Bell Collection at the University of Minnesota Library, is accepted as the globe gores to which allusions were made in the *Cosmographiae Introductio*. The South American region bears the name "America."[6]

The Lenox globe, which is thought to have been made as early as 1510, is generally referred to as the oldest extant post-Columbian globe. Purchased by the great collector of early Americana James Lenox in 1850, it now resides in the New York Public Library. It is an engraved copper ball, about thirteen centimeters (five inches) in diameter, and it depicts an island to the north that is suggestive of Newfoundland but that shows no northern continent. Only "Isabel" and "spagnola" are displayed in the Caribbean; "Zipangri" (Japan) is just north of the northwestern corner of the southern continent, which is insular and contains the names of "Terra de Brazil," "Mundus Novus," and "Terra Sanctas Crucis."[7]

Another globe in the Hauslab collection was a 37-centimeter (14.5-inch) wooden globe that was covered with a preparation to allow the application of paint for a map. It is generally thought to have been produced in 1515. The northern section of the New World is called "Par(ias)," the last three letters having been obliterated; the southern section clearly bears the name "America."[8]

The so-called Paris Green Globe, assigned a date of circa 1515, is located in the Bibliothèque Nationale. It is an undated wooden sphere, 24 centimeters (9.5 inches) in diameter, and displays the name "America" four times, twice in North America and twice in South America—the first time the name appears on both continents.

Another globe and gore map is also given the date of 1515 because the cartographer published an accompanying tract: *Luculentissima quaed terrae totius descriptio . . . : cum globis cosmogaphicis: Noribergae 1515* (A most luminous description of the whole earth . . . with the cosmographic globe 1515). The cartographer was Johann Schöner of Nuremberg. His name provides a logical transition between the above-mentioned globes that suggested the existence of the Waldseemüller world map of 1507 before its discovery and a consideration of the events over the almost four centuries during which that map remained undiscovered.

Johann Schöner (fig. 10) was a mathematician, astronomer, and geographer who was born in 1477 in Carlstadt, Franconia. After holding a church office in Bamberg, he assumed the position of a professor of mathematics in the gymnasium of Nuremberg, where he remained until his death in 1547. He is regarded to have been one of the leading German scientists of his time, and, in addition to his many publications, his globe making earned him a position of prominence.

Schöner assembled and bound the folio in which the Waldseemüller world map was found. Strips of a parchment-print of one of Schöner's terrestrial globe gores were employed as folds of the binding

Figure 10. Johann Schöner (1477–1547).

of the codex that was discovered by Joseph Fischer in Wolfegg Castle in 1901. They were detached by the discoverer, "and the tableaux constructed therefrom were incorporated in the collective volume on rebinding the folio." At the time of discovery, the ex libris of Johann Schöner was noted on the inside of the front cover of the codex that contained the Waldseemüller world map of 1507 and its companion documents.[9]

Before Joseph Fischer discovered Waldseemüller's map at Wolfegg Castle in 1901, all attempts to find the world map of 1507 met with failure, and no one could find a clue that would suggest a possible location. The point of production continued to evoke speculation. The journey the map took from the time it came off a printing press until it arrived at the point of discovery cannot be determined with complete certainty.

In tracing the course of America's Baptismal Document, it is generally believed that, because of its size and complexity, the Waldseemüller map was not printed at Saint Dié but rather at an established press, probably that of Johann Grüninger, in nearby Strasbourg, only forty miles away. As Joseph Fischer wrote:

> The technical execution as well as the pictorial decoration of the map of 1507 presupposes a wood-engraving established fitted up on a large scale and the cooperation of skilled xylographers and eminent artists—conditions, which can be shown to have been realized at that time at Strassburg, but not at St. Dié. At Strassburg towards the end of the 15th and the beginning of the 16th century, the technique of wood-engraving and book-illustration was in its prime, at first with a distinct leaning towards the Schongauer school, afterwards under the influence of Dürer and Burgkmair. The Strassburg printers, such as Knoblauch, Hupfuff, Scott, Grieninger [Grüninger] worked with zeal and farsighted enterprise in the line of book-illustration and single-leaf printing. Particular active was Joh. Grieninger [Grüninger], with whose printing establishment a large wood-engraving department seems to have been connected, which not only supplied the excessive demand of its owner, but other

printing houses likewise. . . . With the Strassburg printers, especially with Grieninger [Grüninger], the printing establishment of St. Dié and Waldseemüller personally had intimate intercourse. . . . On the Carta itineraria Europae of the year 1511 [made by Waldseemüller] and in the accompanying text Joh. Grieninger [Grüninger] is named as printer.

The water-mark of the Waldseemüller map of the world of 1507 points to Strassburg. It appears precisely in the maps of the Strassburg Ptolemy of 1513, which were for the most part drawn by Waldseemüller.

In view of these facts it is highly probable, that the woodcuts for the large map of the world and also for Waldseemüller's globe of 1507 were engraved at Strassburg. The printing itself may then, . . . have been done at St. Dié.[10]

From Strasbourg or Saint Dié, the map can be traced to Johann Schöner, but there is no record of either the date he acquired the Waldseemüller map or the subsequent course of the volume, which bears Schöner's bookplate and its contents consisting of the Waldseemüller map, the portions of Schöner's terrestrial globe gores of 1515, the star chart of Stabius-Heinfogel drawn by Albrecht Dürer in 1515, the "Carta Marina" of 1516, and Schöner's celestial globe gores of 1517.

Schöner's binding with its enclosed documents next appeared as a possession of a Prince Waldburg-Wolfegg-Waldsee at Wolfegg Castle in Baden-Württemberg, Germany, about 115 miles from Strasbourg and about 140 miles from Saint Dié. It is an interesting coincidence that the first two syllables, *Waldsee*, of Waldseemüller's name constitute part of the name denoting the domain of the prince who acquired the map and the family, which has had the map in its possession for almost four hundred years.

Wolfegg Castle (fig. 11) is located in the middle of the moraine of the upper Swabian Lower Alps region in the Ravensburg district of Baden-Württemberg. It is about forty kilometers (twenty-five miles) north-northeast of the city of Lindau, which is situated at the east end of the Bodensee and about twenty kilometers due east of the city of

Weingarten. A drive to Wolfegg from any direction courses uphill to a height of 767 meters (2,500 feet) above sea level.[11]

The town of Wolfegg, which took its name from the castle, lies south of the castle; a relationship between the town, containing buildings that had been erected for officials, and the castle is apparent. The surrounding buildings, once used for agriculture, have given way to stables and an automobile museum. The approach to the castle from the town leads to the choir loft of the parish church, which was built in the baroque style in 1733–1736. The castle is connected to the church by a passageway.

The castle's name comes, in part, from its location. *Egg* derives from *eck*, meaning "corner" or "angle," referring to its position on a protruding mountain ridge. *Wolf* is probably a contraction of the surname *Wolfrans* under which the location was first entered in *Liber decimationis*, the parish's tithing book of the Konstanz diocese in 1275. A small hamlet, called Eigileswilare, existed in the region in the fourth century, and the local parish church was mentioned in 861, making it one of the oldest Christian places of worship in Upper Swabia. Roman and Gothic frescoes have been uncovered in the church. Evidence of even earlier inhabitants is provided by a Celtic grave from the first half of the fifth century BCE and an Alemannic burial ground that was uncovered in the region.

In the twelfth and thirteenth centuries, the "Lords of Wolfegg" were named the owners of a castle. This original lineage terminated, and the castle was transferred to the patronage of Ravensburg. In the thirteenth century, Wolfegg passed to the Truchess family, the oldest line of the High Stewards of Waldburg, which was a name agreed upon by the family at that time and assigned to the family seat located eight miles south of Wolfegg. The castle remains in the hands of the Waldburg-Wolfegg family, which received its royal title from Emperor Franz II in 1803.

Initially, the castle was located farther north on a mountain slope, where remains can still be found. In the fifteenth century, Hans Truchess of Waldburg—Earl of Sonnenberg—expanded that castle,

Figure 11. Wolfegg Castle, modern photograph. Courtesy of Verlag Schnell & Steiner, GMBH & Co., Munich and Zurich.

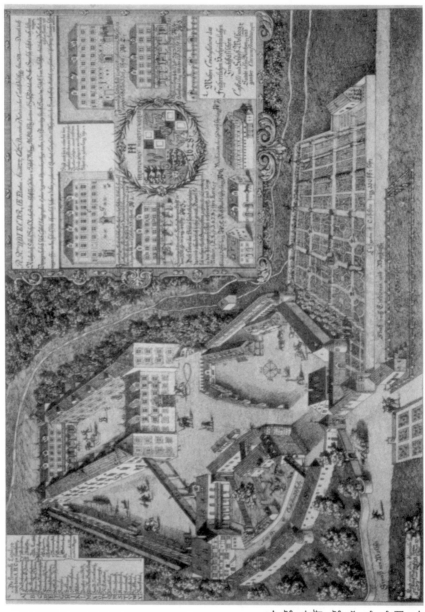

Figure 12. Wolfegg Castle, 1628. Courtesy of Verlag Schnell & Steiner, GMBH & Co., Munich and Zurich.

which was destroyed by a chimney fire in 1578. This served as a stimulus for Jakob Truschess, Baron of Waldburg, to initiate the process of building the current castle in the Renaissance style (fig. 12). The consistency of the four-wing structure presents the best representation of an ideal Renaissance design in southwestern Germany.

Though the basic four-wing structure survived an attack by troops under the command of the Swedish General Wrangel on December 28, 1646, the castle was looted, burned, and gutted. This occurred toward the end of the Thirty Years' War, during which Maximilian Willibald, Earl of Wolfegg, was the field marshal lieutenant who held off the Swedes at the Bodensee and repelled them as they marched against Vienna. Maximilian Willibald had the castle restored, based on the original design, and it was during that period that the folio containing the Waldseemüller world map of 1507 was most likely added to the castle's collection. The folio containing the map, however, does not appear on an inventory until the eighteenth century.

At the end of the seventeenth century, the castle's baroque top floor was built, but the hall of knights was not completed until the eighteenth century. Among the statues lining the walls of the hall of knights is an unflattering representation of Max Willibald, Imperial High Steward, Baron of Wolfegg (1604–1667). The Waldseemüller map and its companions within the binding, during their residence at the castle, were reached by climbing an Italian-French–style broad stairway. This stairway permitted ascension on horseback, to a room on the second floor where the folio was stored in a large old safe with six keyholes. That room, which still contains the safe, is currently the office of the curator of the castle's collection and the location of most of the castle's archival material.

The word *serendipity* derives from the legend of the three princes of Seredip, which later was known as Ceylon and is now Sri Lanka. The princely trio set out seeking one treasure of gems and found an even more valuable treasure trove. Such was the scenario of discovery of the long sought after treasure of the cartography of America, the Waldseemüller world map of 1507.

At the beginning of the twentieth century, Joseph Fischer, a Jesuit priest with the academic rank of professor at the University of Feldkirch in Austria, serendipitously discovered the Waldseemüller world map of 1507. While conducting research on the Norse voyages, which occurred in the period from about CE 1000 to 1300 at Wolfegg Castle, Father Fischer came upon a volume with a strong cover of red beechwood and a banded back of hog skin. The lateral portions of the leather were pressed with Renaissance ornaments consisting of a series of flowers framed by straight lines. Four raised bands appeared along the spine. Pasted on the inside of the front cover was the ex libris of Johann Schöner. The binding contained the once-folded twelve sheets that constitute Waldseemüller's map. In addition, the volume contained the once-folded twelve sheets of the 1516 "Carta Marina," the 1515 star map of Stabius-Heinfogel drawn by Dürer, and gores of a 1517 celestial globe by Schöner. The binding was secured with two strapped brass clasps in Gothic style. When the volume was later taken apart to make facsimiles, strips of parchment prints of Schöner's terrestrial globe gores of 1515 were found. They had been used as folds to bolster the binding.

The discovery of the long sought after map was announced in an article by Professor von Wieser that appeared in the December 1901 issue of *Petermann's Mittheilungen*, a German journal. Three months later, on March 2, 1902, the *New York Times* brought the discovery to the attention of an American readership. It declared: "There has lately been made in Europe one of the most remarkable discoveries in the history of cartography, and one which must hereafter be referred to in all works pretending to give in detailed form the story of the New World."

The monumental discovery is an example of an aphorism credited to Louis Pasteur, the great French scientist, who said: "Le hazard ne favorise que les esprits préparés" (Chance only favors the prepared mind). It was the scholarly background of Father Fischer that made him eminently prepared to recognize his extraordinary discovery, the holy grail of American cartography. Although his presence at Wolfegg

Figure 13. Father Joseph Fischer, SJ.

Castle was stimulated by the hope that the castle's notable library would provide material applicable to the research he was conducting on the Norse voyages and also on Ptolemaic cartography, he recognized the importance of the contents of the volume he had uncovered. He rapidly concluded that among the contents was the Waldseemüller world map of 1507, which had been referred to in the *Cosmographiae Introductio*.

Father Joseph Fischer and his mentor, Franz Ritter von Wieser, set forth several arguments to support that conclusion. They noted: (1) the agreement of the title of the map and the expression "Tipus orbis generalis" in the legend on sheet 5 of the map with statements appearing in *Cosmographiae Introductio*; (2) the fact that the map is composed of several plates, or "tabulis," consistent with the narrative of chapter IX of *Cosmographiae Introductio* in which "tabulis" appears and several plates are alluded to; (3) the agreement between the discovered map and the manuscripts drawn by Glareanus referring to the Waldseemüller map; (4) the inclusion of coats of arms to denote specific realms, and crosses to denote dangerous waters as enunciated in *Cosmographiae Introductio* on the verso of page iiij; (5) the presence of the name "America" on the map as it appears in the narrative of chapter IX and in the margin of the same page in *Cosmographiae Introductio*; (6) the word-for-word agreement between the narrative and some of the legends on the map; and (7) the dominant portrait of Amerigo Vespucci and its confrontation with the portrait of Ptolemy in the map's upper border. Furthermore, the fact that an identical number of sheets was used for the 1516 map found in the same volume and that Waldseemüller's name appeared twice on the 1516 "Carta Marina," which makes direct reference to the 1507 map in a legend, offers reinforcement that the 1507 world map called for on the title page of *Cosmographiae Introductio* had been discovered.[12]

Little has been published about the man who made this monumental discovery. The most complete biography of Father Joseph Fischer is a typewritten manuscript prepared by Wilfried Haller around 1980. This work is preserved among Fischer's papers in the Archivum Monacense Societatis Jesu at the Ignatiushaus in Munich.

Joseph Fischer was born on March 19, 1858, in Quadrath, in the governmental district of Cologne. His father, Gustav, a painter of decorations, enrolled him at the gymnasium in Bergheim, Muenstereiffel, Aachen and Rhine, from which Joseph graduated in 1878. Ironically, his worst grades were in history and geography. That year, he entered the University of Muenster with a plan to study philosophy, but left after only one semester in order to attend a seminary in Cologne; he joined the Society of Jesus in 1881. After spending five years studying theology and philosophy in Exaten, Blyenbeck, and Wyansrad, he was qualified to teach the lower grade at the Stella Matutina College in Feldkirch, Austria. On August 31, 1890, Joseph Fischer received ordination as a priest.

Fischer continued studies in history and geography and, in 1895, passed his state examination in those subjects. That same year he became an Austrian citizen. After a probationary year, he was awarded a "Certificate of Suitability" and would spend thirty years at Stella Matutina teaching history and geography. While on the faculty of that institution, he engaged in research in the field of historical cartography, with a particular interest in Ptolemaic contributions and the Norse voyages. It was in an effort to further his studies in those fields that Fischer initially sought access to the historical collection of the castle. It was probably because the children of Count Waldburg-Wolfegg were studying with Fischer at Feldkirch in the beginning of the twentieth century that Fischer was granted permission to investigate the castle's collection.

The initiative for Fischer's research can be traced to his presence at a seminar conducted by the professor of geography, Franz Ritter von Wieser, at Innsbruck in 1894. The focus of the seminar was the analysis of a large peninsula depicted on some of the early maps of northern Europe and the issue of whether that peninsula represented the earliest delineation of Greenland. During the discussions, Professor von Wieser encouraged Fischer to investigate and write a scientific paper about the discoveries of the Norsemen in America. The so-called Vínland voyages were credited to the leadership of Leif Eriksson and occurred during the early part of the eleventh century.

Fischer proceeded with his research at Wolfegg Castle because of its reputation as a repository of an impressive collection of historical documents. At that venue, Fischer made a series of important discoveries. The stimulus for Joseph Fischer's memorable visit to Wolfegg Castle that resulted in the discovery is detailed in his book on the Norse voyages:

> I must also thank Father H. Hafner, S. J., who kindly consented to search for Ptolemy MSS. In Wolfegg Castle, belonging to Prince Waldburg-Wolfegg, and was fortunate enough to discover a valuable MS. By Donnus Nicolaus Germanus, whose maps of Greenland . . . corresponded to those of the Ulm editions of Ptolemy of 1482 and 1486. I paid a visit to Wolfegg Castle to determine the relation between the Wolfegg. MS. And the printed Ulm editions. Last, but not least, came a most important discovery: the long lost large World Map and "Carta marina" of the cartographer Martin Waldseemüller (Ilacomilus), 1507 and 1516, covering some 24 folio sheets. The lucky discovery was remarkable, if only for its bearing on the discoveries made by the Norse men . . . as well as on their relation to the later discoveries of Columbus and his successors.[13]

Old manuscripts in the castle's collection documented the early representation of Greenland north of Europe. In the summer of 1900, Fischer analyzed a manuscript written by Donnus Nicolaus Germanus, the Benedictine humanist, cartographer, and printer, who added fine modern maps to Ptolemy's *Geographia* when he edited the Ulm editions of 1482 and 1486. Fischer offered proof that the manuscript was the model for the Ulm editions. Shortly thereafter, Fischer discovered the binding that contained three treasures—the Waldseemüller world map of 1507, the 1516 "Carta Marina" by Waldseemüller, and the 1515 Stabius-Heinfogel star chart drawn by Albrecht Dürer. Fischer expressed his belief that the star chart drawn by the renowned artist was the stimulus for the purchase of the entire volume in the sixteenth or seventeenth century.[14]

Joseph Fischer's scholarship allowed him to conclude that he discovered the "lost" 1507 world map by Martin Waldseemüller. He also identified the "Carta Marina" of 1516 and reported his discoveries together with Professor von Wieser in 1903 in *The Oldest Map with the Name America of the Year 1507 and the "Carta marina" of 1516 by M. Waldseemüller*. The work was dedicated to His Highness Prince Franz von Waldburg zu Wolfegg-Waldsee, who was the owner of Wolfegg Castle at the time of discovery. Yet another reward resulted from Fischer's research at Wolfegg Castle. In the attic, he uncovered a world map produced by Jodocus Hondius in 1611. It surpasses the 1507 Waldseemüller world map in size and depicts the progress that had ensued in the century between the construction of the two maps. In 1902, Fischer published his major opus: *The Discoveries of the Norsemen in America with Special Relation to Their Early Cartographical Representation*.

In the winter semester of 1903 to 1904, Joseph Fischer was awarded a scholarship for research by the Istituto Austriatica that allowed him to visit libraries in Milan, Rome, Naples, Florence, and the Vatican. At the Ambrosiana Library of Milan, he was received by chief librarian Dr. Achille Ratti, who later became prefect of the Vatican Library and, ultimately, Pope Pius XI. At the Vatican Library, in the *Codex Vaticanus Latinus*, Fischer discovered a series of different city vignettes. Fischer also defined the importance of the *Codex Urbinas greacus LXXIII* in the Vatican and proved that it was the prototype of many manuscripts in Florence, Venice, Vienna, and Paris. Fischer continued his pilgrimage by traveling to England and France. At the British Museum, he found the *Codex Athous graecus* that had been produced at the Vatopedi Cloister of Mt. Athos. In Nancy, he discovered the *Codex Nancieanus latinus 442*, a map by Claudius Clavus, which Fischer proposed was the earliest map to depict Greenland.[15]

Between 1914 and 1939, Fischer devoted most of his time to research on Ptolemaic geography, concentrating on the interdependence of the maps, their sources, and influences. A series of published essays disseminated his findings, including one analyzing Johann Ruysch's world map of 1508. Shortly before he died, Fischer completed research on the oldest map of Abyssinia to the extent that a colleague could finish the work. That manuscript map of Abyssinia is dated 1456 and appears in the *Paris Codex latinus 4802*. The influence of the map of Abyssinia is manifest on the Contarini-Rosselli world map of 1506, which is located in the British Library.[16]

During his life, Joseph Fischer was a member of numerous esteemed organizations, including the Academy of Sciences in Vienna, the Papal Archaeological Academy, the Royal Geographical Society (1925), and the Philosophical-Historical Society of the Academy of Sciences of Vienna. He was also the recipient of many accolades, including the Gold Medal of S. H. Pope Pius XI (1932), the Silver Karl-Ritter Medal of the Geographical Society in Berlin (1933), a fellowship in the American Geographical Society (1935), and a Doctor Honoris Causa degree of the philosophical faculty of the University of Innsbruck (1935).

In 1929, Fischer retired from teaching but continued to devote his time to research and activities in the ministry and he remained the custodian of various scholarly collections. After 1933, with the advent of the Nazi regime and its anti-Catholic policy (in addition to its anti-Jewish crusade), the College at Stella Matutina was closed, and the building was sold in 1938. Fischer moved to Munich in 1939, and, two years later, Duke Maximilian Waldburg-Wolfegg invited him to return to Wolfegg Castle, the site of Fischer's greatest triumph, to continue his cartographic research.

While in residence at Wolfegg, Fischer served as a spiritual adviser for the duke and Prince Johann. Joseph Fischer died in Wolfegg on October 26, 1944, and was buried at the cemetery of Wolfegg in a grave alongside two other Jesuit priests. In an obituary delivered on the occasion of Joseph Fischer's hundredth birthday, Count Johann Waldburg-Wolfegg offered a personal insight into the life of the priest-scholar, stressing the rare combination of kindness, intellect, humor, and a unique capability of relating to his students.

There is a speculative and intriguing association between Fischer and another cartographic treasure, the famous and notorious Vínland map that Paul Mellon donated to Yale University in 1965 (see plate

12). At the time the map was donated, there was general agreement that Norsemen, sailing from Greenland in the early part of the eleventh century, reached and temporarily settled on North American soil. Currently, the ruin site at L'Anse aux Meadows is the only firm evidence of Norse dwellings in North America.

The Vínland map, which depicts the islands of Vínland and Markland in the northern part of the Western Hemisphere, purportedly provides cartographic evidence of a Norse presence in that region. The provenance of that map has been appropriately referred to as a "black hole." The nebulous provenance is most difficult to reconcile when one appreciates that the map and the two manuscript volumes associated with it were purchased by the Yale donor for about $1 million and have been insured for $24 to $25 million.[17]

From the time of the announcement of its existence to this day, the authenticity of the Vínland map has been the subject of an ongoing debate and several conferences. Kirsten A. Seaver, author of an authoritative book about the Norse exploration of North America, suggests that Joseph Fischer constructed the map, which he presented as a subtle private taunting of the Nazi regime.[18]

The postulated scenario detailing the genesis of the Vínland map, as Seaver convincingly constructs, offers an example of a scholar employing the sword of intellect to strike a blow against a brutish and repugnant antagonist. Fischer, an acknowledged authority on the Norse voyages and Seaver's suggested author of the Vínland map, had been convinced that the Norse conveyed information to the Europeans regarding the existence of an American mainland long before the Renaissance. Fischer also believed that cartographic evidence of the Norse voyages to America must exist and awaited its discovery.

As Seaver deduces, the profile of the cartographer of the Vínland map incorporates several characteristics and requisites. The map manifests evidence of a knowledge of Latin, an appreciation of fifteenth-century cartography, a scholar's knowledge of the Norse discovery of Vínland, and an interest in the early attempts to spread Roman Catholicism. Also, the inclusion on the Vínland map of the

Samoyeds—"Magnum mare tartorum"—and reference to the thirteenth-century journey of Carpini to the Mongols, which probably had as its source a document written in German, suggest that the map was drawn by a German. Joseph Fischer satisfied all these characteristics and requisites. A remarkable fit!

The possibility that the Vínland map is a forgery has been adhered to by many scholars. Kirsten Seaver offers a poignant argument that Father Joseph Fischer was the forger.

> One or more of the map's singularities might separately reflect views held by a well known scholar. However, within the period of 1923–57 there was one person with an educated interest in the medieval Norse[men] and their activities in Greenland and North America who was equally knowledgeable about medieval missionary activities to the far reaches of the known world, and who also was an expert on fifteenth-century maps and cosmography. That one person was known for his passionate interest in all three areas of research. He also understood . . . as Painter and Skelton evidently did not . . . the subtler ramifications of medieval geographical concepts. Dating all the way back to 1440, the only person in the world who could have made the Vínland Map is Father Josef Fischer, S.J., a German/ Austrian contemporary of Luka Jelic.[19]
>
> [Luka Jelic, a Franciscan historian with an interest in the Roman Catholic influence on the Norse in Greenland and America, early in the Vínland map debate, was considered as a possible author of the map or its legends.][20]

Seaver provides evidence to connect Fischer with the name "isolanda Ibernica" that appears on the Vínland map as an indication of a pre-Norse evangelical effort of Irish monks. Another deduced association is related to the suggestion that the Vínland map's depiction and placement of the Vínland island bears a similarity to the representations on the Cantino and Caveri manuscript maps. The latter maps had been intensely studied by Fischer as he sought to determine the role of precursor charts as influences for the Waldseemüller world map of 1507. All three maps include a similar island in the northwestern Atlantic Ocean, northeast of the New World. Another connection between Joseph Fischer and the Vínland map can be extrapolated from Waldseemüller's 1516 "Carta Marina," which the scholarly priest also discovered and intensely analyzed. The "Carta Marina" is the earliest printed map to mention the "Samoyeds," who are also referred to twice on the Vínland map.[21]

The Vínland map is an expression of knowledge of the Norsemen about Greenland and North America after Claudius Clavus disseminated information about them to the Europeans. Claudius Clavus was another focus of Fischer's cartographic scholarship. Fischer's analysis spanned the evolution from a map by Clavus, which included no American island or mainland, to Waldseemüller's map of 1507 and the "Carta Marina," using the maps of Germanus, Cantino, and Caveri as sequential steps.

The conclusion of Kirsten Seaver that Joseph Fischer forged the Vínland map results from sleuthlike scholarship of the document that has been the center of inquiry and argument for a half century. The analysis and deductive reasoning that led to Seaver's thesis requires an appreciation of the documentary elements of the Vínland map as well as the history of the discovery of that map.

Three manuscripts play a role in reconstructing the proposed scenario. In the middle of the twentieth century, two manuscripts, which were bound in a slim volume, were discovered. One was a map of the world including Iceland, Greenland, and Vínland; the other, a narrative known as the "Tartar Relation," was a previously unknown account of the mission of John de Plano Carpini to the Mongols in 1245 to 1247. Both the map and the text appeared to be done by the same hand and were judged to have been constructed in the middle of the fifteenth century, thereby antedating both Cabot's and Columbus's discoveries. The wormholes on the two documents, however, did not match. The third element, a manuscript, discovered and offered for sale separately, is a portion of Vincent of Beauvais's *Speculum Historiale* written in the same hand as the other two manuscripts. Scholars, however, suggested that the writing on the map was made by "a later and much more skilled hand." When the positions of the wormholes on the pages of the three documents were matched, it became apparent that the map had originally been in the front of the *Speculum Historiale* and the "Tartar Relation" at the back.[22]

The copy of the *Speculum Historiale* was incomplete, consisting of only the first four sections. Seaver found that a Swiss auction catalog of 1934 listed what appears to be the missing fifth section and that the section had come from the Mikulov Castle library in Brno. The collection of the castle was well known to Fischer, who had served as a consultant to the curatorial staff of that library. Seaver speculates that when the library was sold, Fischer acquired a bound edition containing the "Tartar Relation" and the *Speculum Historiale*. The use of some of the parchment provided the opportunity to create the Vínland map on period paper and perpetrate a hoax that satisfied Fischer's antipathy toward the Nazi regime.

According to Seaver, the possible motive for Fischer's production of the forgery was his concern that Norse history, a passion of his, had been used as Nazi propaganda for establishing Germanic Aryan superiority. In making the Vínland map, on the one hand, Fischer created a document that, if it fell into their hands, presented for Nazi propagandists a desirable egocentric identification with an Aryan people who were characterized by courage and territorial ambitions, not unlike the aims for expansion of the National Socialist Party of Germany. In the minds of the Nazis, the Norse could be considered racial relatives of modern Germans, and, as such, bestowed on Germany ancestral rights to North America. But, offsetting this appeal is the inclusion on the map of a specific reference to the Catholicism that the Nazis deemed an anathema. One legend on the map refers to the discovery of "Vínland" by "the companions Bjarni and Leif Eiriksson" and also to a trip, shortly thereafter, by Eirik Gnupson, "legate of the Apostolic See and the bishop of Greenland," which credits the Catholic Church with an evangelical heroic role in pre-Columbian America in the face of climatic and cultural adversities. Thus, the map was offered as a tease or conundrum to the Nazis. The Roman Catholic Church was shown to be in America during the same time frame that the Third Reich would have hoped to claim the region based on Aryan voyages. This religious reference would offset enthusiasm for the use of the map for propaganda purposes.

In tracing the course of the Vínland map from the time of its alleged production by Joseph Fischer to the point of discovery by the European dealer who uncovered it, it can be imagined that when Stella Matutina was closed in September 1938, and, after Joseph Fischer and his colleagues left in early 1939, the contents were dispersed. At some point, Fischer's cartographic statement could have been included in looted material, which eventually found its way to a dealer. The obscure provenance of the Vínland map and "Tartar Relation" is certainly compatible with such a sequence. This scenario would satisfy those who have deemed the map to be a forgery.

Moreover, the evidence offered by those doubting the authenticity of the map also focuses on the ink used for the drawing. Walter McCrone, an analytic chemist, used polarized light microscopy to find anatase (titanium dioxide) in the ink in a form that was available only after 1920. Fischer would have drawn the map sometime after 1938, when the Nazi Party closed down the Catholic institution at which he worked.

An alliterative summation of the heroic scholarly cleric who uncovered and recognized the holy grail of American cartography would state that Father Fischer was, perhaps, a forger of one famous map but surely the finder of two other famous maps, the 1516 "Carta Marina" and the Waldseemüller world map of 1507.

CHAPTER 6

THE MAP AND ITS MATES

Maps break down our inhibition, stimulate our glands, stir our imagination, loose our tongues. The map speaks across the barrier of language: it is sometimes claimed as the language of geography.

Carl O. Sauer

The discovery of the cartographic treasure, which included the only known copies of both the Waldseemüller world map of 1507 and Waldseemüller's "Carta Marina," generated immediate excitement and exuberance on the part of the discoverer, Joseph Fischer. As to be expected, this immediate emotional response was followed by a concentrated, in-depth analysis of the find. Professor Fischer was joined in this pursuit by his mentor, Franz Ritter von Wieser. The result of their scholarship was published in 1903 in a folio-sized book, *Die Älteste Karte mit dem Namen Amerika aus dem Jahre 1507 und die Carta Marina aus dem Jahre 1516 des M. Waldseemüller (Ilacomilus)* [The Oldest Map with the Name America of the Year 1507 and the Carta Marina of 1516 by M. Waldseemüller (Ilacomilus)].

The publication describes the content of the collective volume that was discovered at Wolfegg Castle. The size and description of the condition of the twenty-four sheets of the two maps are indicated, as are the watermarks born by each of the printed sheets and the one tracing sheet that was included. Watermarks are created as the emulsion used to make paper solidifies. The watermarks are detected by holding the paper up to the light. The authors suggested that the preserved examples are "not clean copies," in that "the black print of the woodcuts overlies a quadratic network drawn in red ink."[1] This finding led the authors to conclude that they had discovered proof sheets for the two Waldseemüller maps.

The publication goes on to offer arguments that the 1507 world map is, in fact, the map called for by the *Cosmographiae Introductio*, and that Martin Waldseemüller was the cartographer. It is pointed out that the two inset maps within the border of the 1507 map represent the source of the 1512 maps by Johannes Stobnicza. The authors describe the "Carta Marina," as well as the cartographic sources on which the two maps by Waldseemüller were based, and discuss the influence of those two maps on subsequent maps.

In 1901, when Joseph Fischer discovered and analyzed the Waldseemüller world map of 1507, his assessment was based on viewing the map with the naked eye and, perhaps, some minor magnification. His expertise was cartography, not printing and typography. Fischer's assertion that the document he discovered was a proof copy of an edition, which, according to a printed statement appearing on the 1516 "Carta Marina" had consisted of one thousand copies, was based on his interpretation that the red network was located *beneath* the printed material.

Most scholars agree that the woodcuts were prepared for press sometime around 1507 when the text was printed at Saint Dié. Because of the size and intricacy of the map, however, it is generally believed that the map was printed at Strasbourg where Waldseemüller is known to have worked before joining the Gymnasium Vosagense. According to Fischer, "The technical execution as well as the pictorial decoration of the map of 1507 presupposes a wood-engraving establishment fitted up on a large scale and the cooperation of skilled xylographers and eminent artists-conditions, which can be shown to have been realized at that time at Strassburg, but not at St. Dié." Additional support for the contention that the map was printed in Strasbourg is the fact that the watermark of the sheets of the 1507 Waldseemüller map also appears frequently on the paper of other Strasbourg printings.[2]

In 1983, in preparation for placing the Waldseemüller world map of 1507 on display at the Smithsonian Institution's National Museum of American History, the opportunity was seized to conduct a modern analysis and expert typographic appraisal by Elizabeth Harris, curator of graphic arts at the museum. She had available for study the sixteenth-century tooled skin binding and its contents consisting of the 1507 world map and the "Carta Marina," the Schöner celestial globe gores, and the fragments of his terrestrial globe gores. The Dürer star chart that was included when the volume was purchased by the sixteenth- or seventeenth-century Prince Waldburg-Wolfegg-Waldsee had been removed and remained at Wolfegg Castle.[3]

Elizabeth Harris's detailed report indicates that both the world map of 1507 and the 1516 "Carta Marina" were printed in black ink from woodcuts. The world map of 1507 does not indicate the name of the cartographer nor does it contain a date of publication, while the "Carta Marina" specifically credits Waldseemüller as its author in two places and presents the date of 1516 in print. Some of the place names and legends on both maps had been cut in the woodblock, while others consisted of printer's type that had been inserted into holes in the block. Grids of red lines appear on both maps, covering the entire "Carta Marina" but only parts of the 1507 world map. Essential to Elizabeth Harris's conclusion is her determination that these red lines were *over rather than under* the black print in contradistinction to Fischer's initial analysis. Harris also points out that it is not logical to assume that the printer drew lines on the paper and superimposed printed matter on them.[4]

The fact that the red lines are positioned over the black print negates Fischer's published assertion that the Waldseemüller maps, which he had discovered, were early printers' proofs. The overlying red lines might have been added at any time. Therefore, the only extant copy of the Waldseemüller world map of 1507 is an impression that could have been made anytime after the woodblock was prepared in 1507. Fischer actually had come to the same conclusion after he published his book about the two Waldseemüller maps. This is attested to in two letters written by him to Princess Sophie in 1904 and 1905 that were discovered by Kirsten Seaver in the archives of Wolfegg Castle. Fischer also informed his mentor and coauthor von Wieser of this conclusion, which prompted the expression of strong disagreement and displeasure on the part of von Wieser.[5]

Elizabeth Harris's analysis of the paper indicates that all twelve sheets of the 1507 world map and eleven of the twelve printed sheets of the 1516 "Carta Marina" bear the same watermark of a crown, as had been described by Joseph Fischer. The one loose sheet of the "Carta Marina" used paper with an anchor as the watermark. In the sixteenth century, paper was made by emulsifying linen or rag and pouring the mixture into a frame. This contained horizontal and vertical wires, known as lay-lines, to serve as a skeleton to which the emulsion would adhere during the drying process. The watermark was fashioned by a piece of wire sewn onto the lay-lines in such a way as to create a design. The identical nature of the watermarks on the 1507 and 1516 maps is surprising given the nine-year interval. Damage to the watermarks on the 1516 map would be inevitable if that paper had been made from a mold that had been in use for nine years; however, such was not the case.

Harris offers several possible explanations. The most likely scenario is that both the 1507 and 1516 Waldseemüller maps were printed at the same time, with 1516 as the earliest date. Alternative but less reasonable explanations are that the paper used for the 1516 map had been stockpiled by the printer or papermaker for nine years, or that, in 1516, new paper was produced with a mold that had been made in 1507 and had little subsequent use.

The printed sheets of the two maps provide an indirect method for analyzing the woodblocks that were used at the time of the printing. The twelve blocks of the 1507 world map were made of horizontal strips of varying widths. There is evidence that these strips had begun to separate before the impression was made. This feature did not pertain to the 1516 "Carta Marina." Elizabeth Harris is of the opinion that the woodblocks used for the 1507 map, which she had available for study, were no longer new at the time of the printing and showed telltale evidence of aging. Furthermore, she concludes that the aging process had been going on for more than seven years when the map in question was printed.[6]

In the four corners of the world map of 1507 are paragraphs of text. The upper two were incorporated into the woodblocks and printed with the corresponding sheet of the map. By contrast, the lower two legends were printed on separate paper and pasted over blank rectangular areas onto the map itself. Harris suggests that the original blocks of the sections of map onto which the labels were pasted had deteriorated with age and would no longer accept type, thereby requiring the legends to be pasted on. The patch on the lower left portion of the main map contains no watermark nor any clue to its origin. But Harris has identified the source for the patch on the lower right portion of the map. The back of the sheet pasted on the 1507 world map contains printed matter that is identical to both the text and type appearing on a page from the 1515 edition of a guide to palm reading, *Ein schönes Buchlin der Kunst Chiromantia* by Andreas Corvus. It was printed by Johann Grüninger at Strasbourg.

It is possible to imagine Elizabeth Harris using the analytic approach of Sherlock Holmes, enhanced by improved techniques for magnification, proceeding in the role of a forensic investigator. Her knowledge of the method of producing type in the sixteenth century was so extensive that she could determine which printer owned the type and the relative age of the specific type. Of all the printers working in the area during the time of production of the Waldseemüller world map of 1507, only Johann Grüninger owned most of the fonts of type used on that map. One of the world map types that he owned did not exist in the form that appears on the map in 1507. The only world map type that Grüninger did not own was the one used for the *Cosmographiae Introductio* editions that came off the press of the Gymnasium Vosagense at Saint Dié in 1507.

The unique "Saint Dié type" appears in the narrative labels of the world map. Harris suggests that, while the main blocks were being cut in Grüninger's shop in Strasbourg, the labels were being set in Saint Dié to be reviewed by the members of the gymnasium, and then sent to Strasbourg to be modified, so that they could be incorporated into the main blocks.

In a detailed study of one specific label, which begins "Unitia est animal" on the world map of 1507, Harris determines that there had been progressive erosion of the font, leading to the conclusion that old type had been employed. The type used on the "Unitia est animal" label does not show the condition that would be anticipated for 1507 and is more in concert with that used on the "Carta Marina" of 1516. In contrast to the world map of 1507, which shows evidence of last-minute changes, the "Carta Marina" is unaltered.[7]

The demonstration of evidence of splits in the woodblocks used for the world map of 1507, indicating aging, and the use of the back of a page, which came from a book printed in 1515 for a label that was pasted on the world map of 1507, in addition to the identical nature of the watermarks of the paper used for the world map and the dated "Carta Marina," cumulatively provide abundant evidence that the two maps were printed at the same time, no earlier than 1516. Thus, the copy of the 1507 Waldseemüller world map that now resides in the Library of Congress was probably not printed in 1507; rather, it represents a product that was produced some time after 1516.

If, as the evidence suggests, the so-called Waldseemüller world map of 1507 was printed in 1516, this should not detract from its importance. It remains the only known copy of an elegant map that imprinted the name "America" on the continental land in the Western Hemisphere. Such a map was called for as an integral component of the *Cosmographiae Introductio*, which has a known date of 1507. Until another copy of the map is discovered and its date of printing defined, the Waldseemüller map merits its designation as the document that put "America" on the map and, as such, America's Birth Certificate or Baptismal Document.

About two decades after Elizabeth Harris's analysis, when the Waldseemüller map was finally received by the Library of Congress on June 27, 2001, the map was deemed to be in excellent condition. The library's conservators under the leadership of Heather Wanser, the senior paper conservator, conducted a detailed analysis of each of the twelve sheets, emphasizing aberrations and flaws.

The map arrived in its sixteenth-century German binding with a tooled leather spine and wood covers held together by two brass clasps. Before any treatment was initiated, an extensive series of photographs was taken. The evaluation of the map included examination with a binocular microscope to view details, examination under infrared and ultraviolet light to reveal aspects not visible with normal lighting, and beta radiography to make an image of the watermark on each sheet. In addition, the pH level of the paper was determined, and samples of the pigment and paper fiber were taken for analysis. The light sensitivity of the red lines, the iron gall ink inscription, and the black printing ink was measured to determine the guidelines to be used for exhibition of the document.[8]

The twelve sheets of the Waldseemüller world map of 1507 were removed from the binding. The conservators softened the adhesive that had been used for attachment by dabbing the affected areas with a small amount of deionized water, using a small brush. Heat was applied by means of a warm tacking iron through a thin blotter. The remaining adhesive was removed with a spatula and moistened cotton swabs. The map, once removed, required relatively little treatment. A few minor tears were repaired with Japanese tissue adhered with wheat starch paste and A4M methylcellulose. A few small areas of missing paper were filled with antique western laid paper.

The map is constructed in such a fashion that there are three rows, each containing four sheets. The top row is made up of sheets 1 through 4; the middle row, sheets 5–8; the bottom row, sheets 9–12. Each sheet has a fold running down the center, creating a left and a right page. Sheet 1 contains a small red spot at "320" on the left page and a small red spot on the verso of the right page. The verso of the left page has the Wolfegg stamp, and there is also a stamp on the back of the right page. The two outer corners and the inside top edge of sheet 2 are repaired. There is an insect hole one millimeter in diameter in the center of the right leaf. There are three tears (.5 mm.–1 mm.) along the bottom of the left page, a one-millimeter tear on the top of the right page and a two-millimeter × one-millimeter loss at the top of the right page of sheet 3. The verso of the right page contains a small black ink spot and a red smear in the same lower location. Sheet 4 has a repaired tear at the center of the bottom edge of the right page and a two- to three-millimeter paper mend on the bottom center of the left page. There is a small insect hole in the center of that page.

On sheet 5, there are two tears at the bottom to the left of the binding and below "310." There is also a tear above "310" and a forty-millimeter puncture mend near the upper left corner. There are two small tears on the right page at the inside bottom. An insect hole is present in the center of the right and left pages. On the left page of sheet 6, there are two repaired parallel tears and a small repaired tear at the inside top edge. The right page has a repaired tear at the inside bottom edge. A red ink spot appears between "50" and "60" and a grid in red ink appears on portions of the sheet. The grid drawn in red ink also appears on parts of both the left and the right pages of sheet 7. There is a repair to the outer border of that sheet from "20" to "30" and a puncture mend to the outer corner near "20." Along the lower edge of the left page, there are three small mends. The upper right corner of sheet 8 has been replaced, and there are three one- to four-millimeter tears along the bottom edge of the right page. The verso of the right page has a tide line running the length of the page and some spotting near the bottom ink print, or mold damage.

Sheet 9 has drawn the greatest attention because it includes the word "America"—the first time this word appeared on a map. On the reverse side of this monumental sheet, the conservators discovered an identical offset image with the name "America" (see plate 13). This was probably made when that sheet was placed on another freshly printed sheet, indicating that at least one additional copy of that sheet existed. The sheet has several small tears along the edge, and an eighty-five-millimeter tear repair running parallel with the bottom edge beginning about eighteen millimeters from the bottom. A printed label is pasted on the page. The outside border of the left page of sheet 10 contains a handwritten notation in ink. There is a small red spot on the right page just above "80." There is also a small red spot on the left page of sheet 11 near "Cirobena." Sheet 12 has a pasted printed label in the lower right corner, a five-millimeter repair of the right page near the gutter, and a 110-millimeter repair along the bottom edge of the left page.

The Waldseemüller world map of 1507 (plate 14) consists of twelve sheets of woodcuts that are engraved on handmade, laid rag paper showing twelve chain lines. The size of the single sheets including the white margin is 45.5 × 62 centimeters (17.5 × 24.5 inches). The impression made by each of the twelve woodcuts is strong. Each of the twelve impressions has a distinct border suggesting that the work was prepared to be displayed in a volume. On the other hand, trimming of the borders allowed for the assembling of a single large map, the size of which exceeded all previous printed maps. It is the earliest printed wall map. With the twelve sheets trimmed to the borders and joined, the resulting map measures 132 × 236 centimeters (approximately 52 × 93 inches, or 34 square feet). Joseph Fischer noted that, when joining the sheets to make a facsimile of the whole map, the match was, at times, off by as much as two centimeters (one inch).[9]

The joined sheets form an upper border and two lateral borders that are uninterrupted. The lower border is broken by the most southern portion of Africa and the water south of that land. Below the lower border is the title "VNIVERSALIS COSMOGRAPHIA SECVNDVM

PTHOLOMÆI TRADITIONEM ET AMERICI VESPVCCI ALIORVQVE LVSTRATIONES" (note the letter *U* is written as a *V*). This translates as: A drawing of the whole earth following the tradition of Ptolemy and the travels of Amerigo Vespucci and others.

The main large map is presented on a modified conical or cordiform (heart-shaped) projection that characterized Ptolemy's productions. It was Ptolemy, in the second century, who introduced the method and names of latitude and longitude. His method of measurement and selection of the Canary Islands as the prime meridian resulted in the elongation of land eastward from this line and consequently reduced the distance, presumably made up of water, between Europe and Asia.[10]

Filling the space between the main map and the work's rectangular border is an elaborate depiction of shaded clouds and wind heads. (In the time of Ptolemy, the winds were personified and assigned distinctive names.) Each of the four corner sheets contains a legend; the two upper legends are integral to the woodcut while the lower two are pasted onto blank boxes. The two middle upper sheets, 2 and 3, contain inset hemispheric maps and the portraits of the men associated with each of the hemispheres of Waldseemüller's world map of 1507.

The border of sheet 1 contains, in its corner, a legend in Latin that states:

> Many have thought it an invention what was spoken by the celebrated poet [i.e., Virgil] that beyond there lies a land past the equator where heaven-bearing *Atlas upon his shoulders turns the wheel of heaven fixed with gleaming stars.* Now, finally, this has become clear. For there is an island, discovered by Columbus, Captain of the Castilian King and Amerigo Vespucci, men of great and surpassing character, which, not withstanding the greater part, lies beneath the equator and between the tropics, nonetheless extends beyond the equator approximately nineteen degrees beyond Capricorn toward the Antarctic pole. In this land much gold and other metals have been discovered. [Italicized text taken verbatim from *Aeneid* 4: 482.]

The border of the sheet includes the wind head of the west wind, "*ZEPHIR*⁹" (Zephyrus), blowing in an easterly direction and, according to the *Cosmographiae Introductio*, "having a mixture of heat and moisture, melts the snow." At the top of the document, the wind "CHOR⁹" (Chorus) "blasts forth in a southerly direction."

The wind heads of "CIRCIUS" and "SEPTENTRIO" appear on sheet 2, and there is a legend in the lower right corner within the sea north of the northernmost part of Europe, east of Greenland, which is depicted as a peninsula. Dominating the sheet is a depiction of "CLAUDII PTHOLOMEI," the Alexandrian cosmographer, who is holding a quadrant and symbolically facing toward the east, viewing the inset map of the so-called Ptolemaic world (fig. 14).

Shown on the inset map, within 180° of specified longitude and extending from 66° north latitude to 40° south latitude, are all of Europe, England, Ireland, the Scandinavian Peninsula, Greenland (shown as a peninsula), and all of Africa with the exception of the most southern portion, which is cut off by the line defining longitude. There is a line across Africa indicating that the southern third of that continent is made up of unknown land. A large sea borders the west coasts of Europe and Africa and is in continuity with the Mediterranean Sea. The vast majority of Asia is included, excluding only the eastern portion. A large body of water is present north of Europe as well as north of western and central Asia. The Arabian Peninsula is appropriately represented, unlike the depiction on the "Carta Marina," while the geography of India is not recognizable, but an Indian Sea is named. Within the Indian Sea there is a large island named "taprobana" (Sri Lanka), and, to the south, are "madagascar" and "znuzabar."

The contents of the border of sheet 3 are certainly the most striking and are critical to designating the document as America's Birth Certificate. Half of the wind head of "APARCTIAS" appears, as does the entirety of the wind head designating the north wind, "AQVILO." According to the text of *Cosmographiae Introductio*, it "by reason of the severity of its cold, freezes the water." Above the banner

Figure 14. Claudii Phtolomei and Ptolemaic World, sheet 2 (top row/second from left), Waldseemüller world map of 1507.

with the name "AQVILO" within the clouds is a single winged insect that appears to be a fly.

It had been speculated that perhaps it was actually a wasp flying behind the head of Vespucci, because the Italian and Latin word for wasp is *vespa*—the root of Vespucci's name and the dominant feature of the family crest. It is more likely that its presence is in keeping with the practice of Netherlander, German, and Italian artists during the latter half of the fifteenth century and the early decades of the six-teenth century. It was common to insert a flying insect as part of a graphic to serve as a "protective talisman against the real insects which might otherwise settle and leave their dirt marks brushwork of a sacred theme" or bore through and create wormholes.[11]

In the border, Vespucci holds a large compass and symbolically faces west, as an antithesis to Ptolemy. The relative importance accredited to Vespucci is apparent in the size of the letters in the banner bearing his name, "AMERICI VESPVCII," which far exceeds that of Ptolemy's printed name. Subject to Amerigo Vespucci's view is an inset map (fig. 15) extending from 70° north latitude to 40° south latitude, and from 190° to 360° longitude. It shows the eastern portion of Asia separated by a large body of water from the west coast of the two continents in the Western Hemisphere. Within that ocean, "Zipangri insula" (Japan) is positioned in an inappropriately short dis-tance from the west coast of Central America. The west coast of North America and Central America runs in a straight line along the 280th longitudinal line. Similarly, the west coast of South America is defined by two straight lines meeting at a slightly obtuse angle.

North America and South America are presented as a continuous landmass, with an east coast showing elements of capes and bays. The two continents are joined by an isthmus. The northern part of the northern continent is abruptly ended with a straight horizontal line at 55° north latitude. The continent is shown with the suggestion of a Florida peninsula, a Gulf Coast, and a Yucatán Peninsula. (This is before Florida was supposedly discovered by Ponce de León in 1513 and before Pineda first charted the Gulf Coast in 1519.) The two names

Figure 15. Amerigo Vespucci and Western Hemisphere, sheet 3 (top row/third from left), Waldseemüller world map of 1507.

appearing on North America are "Caput de bona ventura" and "caput doffin de abul." Although several smaller islands are shown, the only Caribbean islands named are "isabella," denoting Cuba, and "spagnolla." They are located in "Oceanus Occidentalis." The southern end of the southern continent is cut off at about 50° south latitude by the line that indicates the measurement of longitude. The large southern landmass is called "Terra incognita" and contains several names along the east coast: "arcay," "Caput descado," "Gorsso fremoso," "Caput S. crucis," "monte fregoso," "Allapego," and "rio decandz."

The border of sheet 4 presents an extension of the clouds and the wind heads of "CECIAS" blowing from the north and "SVB-SOLANVS" blowing from the northeast, and also a legend printed from the woodblock that states: "In describing the general form of the world, it seemed proper to set out the inventions of the ancients and those things that have been discovered since then by the moderns. As students of such matters desire to know diverse things, we have, by our labors, provided everything promised, matters randomly known as well as those brought to light diligently and clearly arranged in a single book."

Sheets 5 and 8 contain borders that balance each other. The banner of the wind head "AFRICUS" on sheet 5 is matched by that of "WLTVRN9 EVRUV." The two central sheets, 6 and 7, have no borders. Among the clouds of sheet 9 are the wind head of "AFRICUS" and the wind head and banner of "LYBONOTH9." The text of the paper that was pasted on the blank rectangle in the corner of sheet 9 states: "A general description of various lands and islands, of which the ancient authorities make no mention, *from the year 1497 to 1504, lands discovered on four sea voyages—twice for Ferdinand of Castile and twice upon the southern sea for his most serene highness our lord Manuel, of Portugal—by Amerigo Vespucci, one of his captains and prefects.* We have carefully provided in this picture a true understanding of the location of many places about which there was no information" (italics mine).

On sheet 10, the rim designating longitude is interrupted to allow for inclusion of the tip of Africa and the water south of that land. This differs from the hemispheric inset map on which the tip is cut off. It also contains half of a wind head blowing due north and its banner, "NOTUS." Sheet 11 contains an uninterrupted line of longitudinal measurements and clouds, with half of the wind head, ("AUSTER") and its banner. Sheet 12's right lower corner contains a continuation of clouds and the wind head of "WLTRURN9" and the wind head and banner of "EVRONOTVS." A rectangle contains a legend that had been pasted on. The legend states:

> Notwithstanding that the ancients were most studious in describing the lands of the earth, many remained unknown even to them. *This was the case with America, called after the name of its discoverer, which is thought to occupy a quarter of the world.* Part faces the meridian of Africa, beginning seven degrees this side of Capricorn, and broadly extends beyond the torrid zone and the [*egoceri?*] tropic to the south. And thus also in the eastern tract is the region of Cataia, and of southern India, located past 180 degrees longitude. What was previously known we have added, in order that whoever admires such things, giving attention to what now lays before our eyes, may praise our diligence. One thing we ask, however, that those who are inexperienced and ignorant of cosmography do not straightaway condemn this before making these things more clear to themselves that afterward they may understand the things to come. (italics mine)

It is this piece of paper that contributes significantly to the argument, based on the printed text on the back, that the map purchased by the Library of Congress from Prince Waldburg-Wolfegg was actually produced around the same time as the 1516 "Carta Marina." Because the printed material on the back was published in 1515, the rectangle could not have been covered before that date.

The main map is spread over all or parts of the twelve sheets. The pseudo-cordiform projection of the geography is similar to that which had been employed by the large wall portolan (manuscript sailing chart) of Henricus Martellus. One example of a manuscript map by

Martellus, dated circa 1490, is included in the collection of the British Library, and a similar but larger copy is in the Yale University Library.

The main presentation of the Waldseemüller map is divided into two halves; the line of division running east of Africa, the Arabian Peninsula, and through the middle of the Caspian Sea. The right half of the map contains all of Asia and shows the east coast of that continent. The northern and eastern portions of Asia are drawn according to the descriptions of Marco Polo. Almost all of the legends replicate Marco Polo's narrative, as do the legends on "Zipangri," "Java major," "Java minor," and the other islands in the "Oceanus orientalis," as well as in the "Oceanus indicus meridionalis." These legends derive from a Latin manuscript, a printed edition of Marco Polo, or a redaction of Marco Polo's writings by Fra Pipino. Although this half of the main map is replete with legends, there is minimal evidence of artistry when compared to the "Carta Marina." No scenes are depicted and not a single regal figure is included. India is misrepresented, and "Taprobana" (Sri Lanka) is misplaced and disproportionally large. "Zipangri" (Japan) appears at the extreme eastern portion of the main map on the upper right corner of sheet 4.[12]

Sheets 2, 6, and 10 of the left half of the map incorporate Ptolemaic sources for most of Europe and Asia. Fischer and von Wieser indicate that Waldseemüller used the 1486 Ulm edition of Ptolemy as his main source for Europe and the map of Donnus Nicolaus Germanus for Scandinavia, ignoring the Greenland peninsula, which had been introduced by Claudius Clavus around 1424–1427 and was included on both the Cantino and the Caveri charts. The banners of several countries appear, but no regal or peasant figure is shown. All of Africa with the exception of the extreme western portion takes up a large segment of the three sheets. The east and west coasts of Africa are marked with emblems of Portugal. The depiction of southern Africa draws from Portuguese sources. A large elephant and a group of natives in the center of the southern portion of Africa are the only mammals to be found on the entire map. The southernmost Cape of Good Hope is defined and named "Caput de bona speranza."[13]

The representation of the New World extends through and dominates sheets 1, 5, and 9. The two large landmasses in the New World are located between two large bodies of water, which separate Europe and Africa to the east and the Orient to the west. The ocean to the east of the New World is named "Oceanus Occidentalis" (Western Sea). That specific name was applied for the first time on this map and on the accompanying globe gores.

On sheet 1 (plate 15) in the Caribbean Sea, the large islands of "SPAGNOLLA INSVULA" (Hispaniola) and "ISABELLA INSVLA" (Cuba) stand out. The other named islands in the area are "babueca, sarmento, Sanra, carij, magna, matubiza, ianucanaca, incaio, tartaga, boriquen, laouizes mil virgium, de gada lupo, de sorana, marigalante, to dos sanctos, iamaiaua, and riqua."

The most striking difference between the depictions of the New World on the inset hemispheric map and the main map is the presence of a strait. It is positioned just north of the equator, between the two continental landmasses on the main map, in contrast to an isthmus joining the continents on the inset map. The isthmus or strait area is not shown on either the Cantino map or the Caveri chart that are regarded to be sources for the Waldseemüller world map of 1507.

Fischer and von Wieser conducted a detailed analysis of the Spanish and Portuguese place names appearing on the 1502 Cantino manuscript and the place names on the Caveri manuscript of circa 1504–1505. They compared those names with the names that appear on the Waldseemüller map of 1507. They noted a greater congruity between the Caveri chart and the Waldseemüller map.[14]

The large landmass north of the strait that separates the two continents in the Western Hemisphere has the emblem of Castile at the northernmost and southernmost points. This indicates that all of the land (North America) was claimed by Spain, as evidenced by the presentation of the emblem of that nation in the region. The land is cut off by a straight line at about 60° north latitude, and below the line is stated: "The land beyond is unknown." Within the variegated west coast, a curved line extends the length of the land, and, to the east of that line, it is also indicated that "[t]he land beyond is unknown." To the west of the line, a mountain range extends the length of the continent, perhaps suggesting the Rocky Mountains. To the south of 30° north latitude, only one word appears on the land, and that word, "PARIAS," is presented in capital letters. The word possibly derives from *pariahs*, because the Europeans regarded the Native Americans as an inferior class.[15]

The east coast depicts a geography that can be equated with the Florida peninsula, Gulf Coast, and Yucatán Peninsula. This is analogous to the preceding Cantino and Caveri charts and presents a major enigma. The major question is, What was the source of this geographic concept, and how does it reconcile with the fact that Florida had not been explored until 1513—when Ponce de León conducted his expedition from Puerto Rico? Also, we currently regard the 1519 manuscript of Pineda as the first to define the Gulf Coast. The east coast of the northern continent is replete with capes, inlets, and bays. Twenty-one Spanish and Portuguese place names are assigned to capes, rivers, lakes, and regions. None of these designated names has persisted.

On the continental land to the south of the strait, shown on sheet 5 (plate 16), only two expressions are presented in capital letters. One reads: "TOTA ISTA PROVINCIA INVENTA EST PER MANDATUM REGIS CASTELLE" (This province was discovered by order of the King of Castile). The other is one word—"AMERICA," which is also present on sheet 9. Both statements are absent on the 1513 Ptolemy map by Waldseemüller, on which the land is called "TERRA INCOGNITA," and on his 1516 "Carta Marina," on which "TERRA NOVA" appears. None of the other place names, which are confined to the east coast, achieved permanence.

The east coast of the southern continent delineates capes and bays, as well as rivers emptying into the sea. The flag of Castile (Spain) is planted on a cape at the middle of the coast, and a legend located above the flag indicates that "[t]hese islands were discovered by the Genoan admiral Columbus by commission of the Castilian king." A large bird named "Rubei pfitari" is shown in the midst of the land. A legend on sheet 9 (plate 17) is positioned south of a vessel coasting off the eastern shore of the southern part of South America. It says that "the vessel was the largest of ten ships, which the King of Portugal sent to Calicut, that first appeared here. The island was believed to be firm[?] and the size of the previously discovered surrounding part was not known. In this place men, women, children and even mothers go about naked. It was to these shores that the King of Castile later ordered voyages to ascertain the facts."

The west coast is represented by straight lines joined at an obtuse angle in the northern half and includes the statement "TERRA ULTRA INCOGNITA" (The land beyond is unknown) just east of the lines. A mountain range suggestive of the Andes is shown to the east of that statement and continues to the southern end. The west coast of the southern third of the continent is irregular, suggesting the existence of capes and inlets.

On sheet 2, in the middle of the sea, is an island with the inscription "litus incognitum" (unknown shores), derived from the voyages of the Corte-Reals to Newfoundland in 1500–1501. On sheet 5, the "Islands [Cape Verde Islands] belonging to the Portuguese in the changing times of Prince Henry, the year 1412" are identified, and obviously have been added to the map after the initial woodcut block was printed.

THE MAP'S MATES

Waldseemüller's Gores for a Globe, 1507

One significant map, which is closely associated with the Waldseemüller world map of 1507, was not present among the items found by Father Joseph Fischer at the Wolfegg Castle. The absent map depicts gores for a globe (plate 18). That map, like the larger, more detailed, and more elegant Waldseemüller map, had been specifically referenced on the title page of the *Cosmographiae Introductio*. It is now known to exist as four examples.

The relationship between the gore map and the large wall map can be regarded as that of fraternal twins. The two dissimilar maps grew from the fertile mind of Martin Waldseemüller in 1507. But, to continue the metaphor, these maps arose from separate ova and therefore shared no similarities other than presenting two continents in the Western Hemisphere and containing the words "America" and "Oceanus Occidental" (Western Ocean). After their birth, the productions were separated. Moreover, no copy of the gores accompanied the large wall map as it found its way to the collection of Johann Schöner and subsequently to Wolfegg Castle.

The gore map, though announced on the title page of *Cosmographiae Introductio* and anticipated like the large world map of 1507, remained unrecognized for almost four centuries. The first of the currently extant copies was discovered toward the end of the nineteenth century. Then, in the middle of the twentieth century, the second and third copies were discovered. More recently a fourth copy has been authenticated.

The first copy of the 1507 globe gores of Martin Waldseemüller was found in 1890 in the collection of Austro-Hungarian field marshal Franz Ritter von Hauslab (1789–1883) in Vienna. Lucien Gallois, at that time, recognized it as a Waldseemüller product. Von Hauslab had a distinct interest in cartography and is credited with introducing surface-color technology to Austrian mapmaking. After von Hauslab's death, the princes of Liechtenstein purchased the gore map and transported it to their castle in Vaduz. In 1954, Prince Johann II of Liechtenstein put the map up for sale. After initially dealing through the distinguished New York bookdealer Hans Peter Kraus, the map was bought through Richard Zinser by John Parker on behalf of the James Ford Bell Collection at the University of Minnesota for approximately $50,000.[16]

It is an interesting coincidence that a map drawn by a man whose name translates as *"miller* of the forest sea" should find its way to the collection of a man, James Ford Bell, whose company Washburn Crosby joined with four other flour-milling companies to form General Mills in 1928. In 1926, Bell purchased Walter Bigges's *A Summarie and True Discourse of Sir Francis Drake's West Indian Voyage*, which was the first book in the library that he would eventually establish at the University of Minnesota in 1953. Now this collection contains over twenty thousand items.[17]

While the map was in the collection of the princes of Liechtenstein, it was reproduced in facsimile for private distribution. A title was added indicating that it was probably the globe of Waldseemüller made to accompany a tract titled *Globus Mundi*, printed in 1509. But Harrisse refutes this association because there is no evidence that Waldseemüller is the author of *Globus Mundi* and there is no connection between the book and the map that can be established. The fact that the word "America" appears on the gores is evidence that the map was produced about the same time as the Waldseemüller world map of 1507.[18]

Yet another coincidence relates to the Hauslab-Liechtenstein-Bell copy of the gore map and the Waldseemüller map. In 1975, the current residence of the Waldseemüller map, the Library of Congress, acquired the Hauslab-Liechtenstein collection of maps that von Hauslab assembled during the nineteenth century. The collection consists of about nine thousand map sheets, representing six thousand titles produced from the sixteenth through the nineteenth century. Also, the Prints and Photographs Division of the Library of Congress has among its holdings the *Hauslab Album*, which depicts 257 original broadsides (large pieces of paper with printed material on one side) of land and sea battles extending from 1566 to 1711.[19]

The second copy of the Waldseemüller gores has an even more dramatic history eventuating in the production of a catalog titled *America: Early Maps of the New World*. The copy is now a treasure of the Bavarian State Library (Bayerische Staatsbibliothek) in Munich and is contained as an insert in a copy of Ptolemy's *Geographia*, which was printed in 1486 in Ulm, Germany, by Johann Reger. The volume's residence can first be traced to the Harz Mountains in north-

PLATE 8. Peter Apian[us], "Tipus Orbis Universalis Iuxta Ptolomei Cosmographi Traditionem Et Americi Vespucii Aliorque Lustrationes A Petro Apiano Leysnico Elucbrat," 1520, Vienna, woodcut on paper, 11 x 16 in. (285 x 410 mm) (excluding print outside border).

PLATE 9. Gerhard Mercator (no title), "Joanni Drosio suo Gerardus Mercator Rupelmudan," 1538, Louvain, copperplate engraving on paper, 13 x 21.5 in. (350 x 550 mm). Courtesy of Rare Book Division, New York Public Library; Astor, Lenox and Tilden.

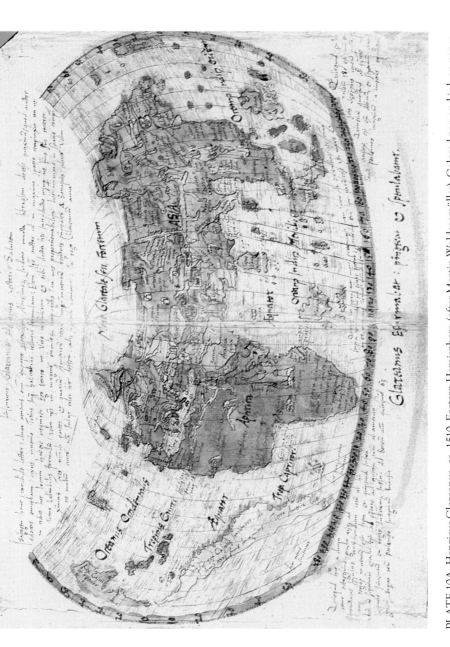

PLATE 10A. Henricus Glareanus, ca. 1510, Eastern Hemisphere (after Martin Waldseemüller). Colored pen and ink on paper, 7.5 x 10.5 in. (195 x 265 mm). Inserted into *Cosmographiae Introductio*. Courtesy of Universitätsbibliothek, Munich.

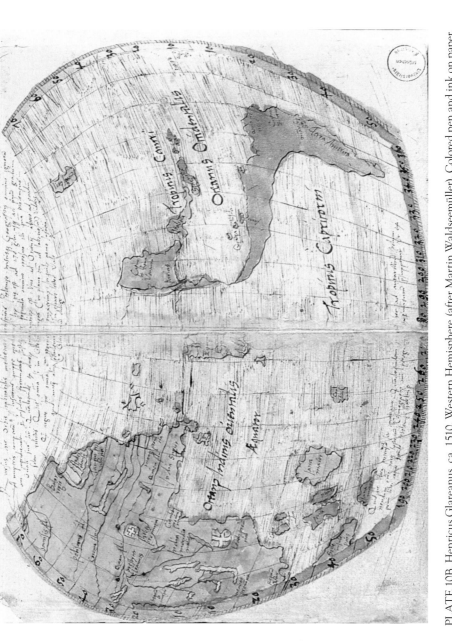

PLATE 10B. Henricus Glareanus, ca. 1510, Western Hemisphere (after Martin Waldseemüller). Colored pen and ink on paper, 7.5 x 10.5 in. (195 x 265 mm). Inserted into *Cosmographiae Introductio*. Courtesy of Universitätsbibliothek, Munich.

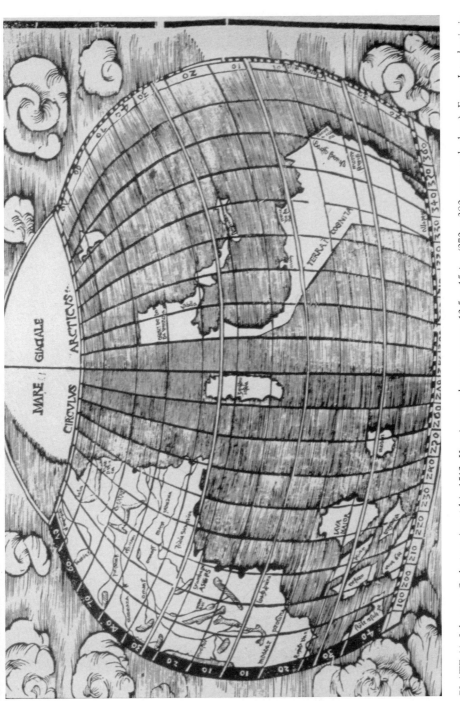

PLATE 11. Johannes Stobnicza (no title), 1512, Kraców, woodcut on paper, 10.5 x 15 in. (270 x 380 mm, each sheet). From *Introductio in Ptholemei Cosmographiam*.

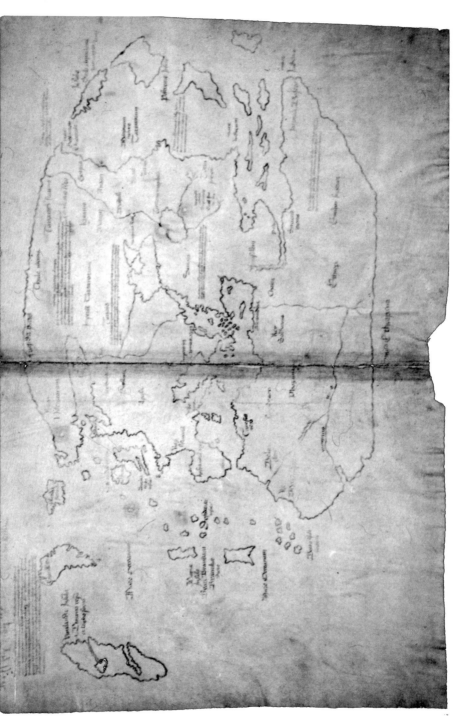

PLATE 12. *Vinland* map, two sheets joined, ink on parchment, 11.25 x 16.5 in. (285 x 420 mm). Courtesy of Yale University, New Haven, Connecticut.

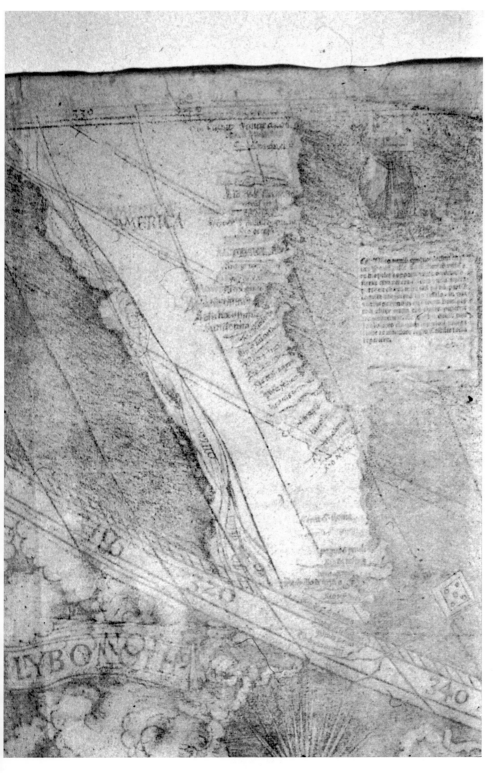

PLATE 13. Verso (backside) Sheet 9 (lower left, bottom row) Waldseemüller world map of 1507. Courtesy of Conservation Department, Library of Congress, Washington, DC.

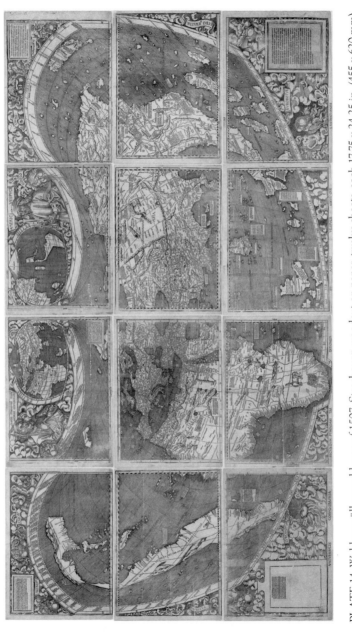

PLATE 14. Waldseemüller world map of 1507, Strasbourg, woodcut on paper, twelve sheets, each 17.75 x 24.25 in. (455 x 620 mm), total when joined 50 x 95 in. (1,320 x 2,360 mm).

central Germany, where, as can be deduced from its eighteenth-century ex libris, it was part of the famous thirty-five thousand–item Wernigerode Castle Library of Count Christian Ernst zu Stolberg-Wernigerode (1691–1771).

According to a second ex libris, the book and presumably the map were part of the nineteenth-century library of Sergei Alexandrovich Soblevski (1803–1970), who collected about twenty-five thousand volumes. His heir, S. N. L'vova, sold most of his library to List and Francke, an antiquarian firm in Leipzig. The book can next be traced to Andreas Joachim Mortimer Graf Maltzan, Freiherr zu Wartenberg und Penzlin (1863–1921) at Militsch Castle, north of Breslau.

The 1486 copy of Ptolemy's *Geographia* with the inserted gores was featured at a London auction conducted by Sotheby's on May 31, 1960, where it was purchased for about $35,000 by the Viennese émigré Hans Peter Kraus (1907–1988). As noted, he was a New York City rare book and map dealer, the same man who was initially involved in the sale of the Hauslab-Liechtenstein gores. In 1990, after Kraus's death, his widow offered the volume for sale in a catalog, at which time the Bavarian State Library purchased the treasure for $1,500,000, providing evidence of the extraordinary financial appreciation of rare maps, particularly those depicting early America.[20]

The third example of the Waldseemüller gores was discovered in the Historischen Bibliothek der Stadtbücherei Offenburg, Baden-Württemberg, Germany, the same state in which Wolfegg Castle is located, about eighty miles to the southeast. Dr. Vera Sack made the discovery in 1993 and described it in an unpublished memoir. By order of the minister of cultural affairs of Baden-Württemberg, Dr. Sack of the Universitätsbibliothek Freiburg and Dr. Gerd Brinkhaus of the Universitätsbibliothek Tübingen had been assigned a restoration project in a new building of the Public Library of Offenburg.[21]

In accordance with the "Reichsdeputationhauptschluss" of 1803, the library was placed in the ownership of the town of Offenburg from 1808. Until recently, it was located in the Grimmelshausen-Gymnasium, where a monastery of the order of "Franzikaner und

Minoriten" had been located previously. In 1997, the library was moved to the Stadtbibliothek Offenburg. Dr. Sack discovered the map inserted in a book, *Aristoteles: Ethica Nicomachea* published by Johann Faber in Freiburg in 1541.[22]

On June 8, 2005, the fourth copy of the gore map was purchased by the English firm of Frodsham and Company, Ltd., at an auction conducted by Christie's in London. The price, including commission, was the equivalent of $1,002,267. The copy differs from the other three in that it consists only of the gores without surrounding blank paper. The provenance is not known. Apparently it was purchased by a European gentleman as a curiosity from a shop sometime in the 1950s or 1960s. Only recently did the owner investigate the item and come to the conclusion that it was of great significance.

When the four examples are compared, the printed material is identical on all copies. The twelve woodcut gores result in a map size of 24 × 39 centimeters (9.8 × 15.4 inches) that could be affixed to a sphere about 12 centimeters (4.5 inches) in diameter. The Old World follows the Ptolemaic representation with the additions of Madagascar, Zanzibar, and Zipangu (Japan), resulting from the narratives of Marco Polo. The representation of the New World includes two large Caribbean islands and two continental landmasses separated by a strait. The northern continent's western portion abuts a meridian designating latitude. The eastern coast depicts a (Florida-like) peninsula and the suggestion of a gulf (of Mexico) and a (Yucatán) peninsula. The east coast of the southern continent is slightly sculpted while the west coast consists of a series of straight lines forming two obtuse angles. The sea to the west is named "Oceanus Occidental" (Western Ocean).

As Fischer and von Wieser point out, the equator on the gores was drawn according to sea charts, whereas on the large Waldseemüller world map of 1507, the equator is drawn as shown on Ptolemaic maps. On the gores, the coast of Guinea is ten degrees closer to the equator than on the large plane map. But there is no difference in the position of the equator relative to the continents between the inset and main representation on the world map of 1507. Also, on the gore map, in the Western Hemisphere, the Tropic of Cancer runs south of Haiti, whereas on the large world map and its inset map, the Tropic of Cancer passes through the middle of the island of Cuba. In the New World section, the parallels on the gores correspond with the contemporary Spanish and Portuguese charts, including the Juan de la Cosa manuscript, the charts by Bartholomew Columbus, the Cantino map, and the Caveri chart. In Africa, however, the location of the parallels makes concessions for the Ptolemaic view. In all other respects, the gores and the Waldseemüller world map of 1507 agree. On this basis, Fischer and von Wieser concluded that the Hauslab-Liechtenstein globe, the only known example at the time of their writing, was the "missing globe of Waldseemüller."[23]

OTHER ITEMS FOUND WITHIN THE BINDING

Unlike the Waldseemüller gore maps that were all eventually located in locales at a distance from the Wolfegg collection, the other items associated with the Waldseemüller world map of 1507 shared space within the binding that Joseph Fischer discovered in Wolfegg Castle in 1901. Considered according to the chronological order of their production, these items are the remaining parts of the terrestrial globe gores of Johann Schöner, the Stabius-Heinfogel star chart engraved by Albrecht Dürer, Martin Waldseemüller's 1516 "Carta Marina," and the gores of a 1517 celestial globe by Schöner.

Parts of 1515 Terrestrial Gores of Johann Schöner

Fischer and von Wieser noted that strips of parchment print of Schöner's terrestrial globe of 1515 had been used as folds when preparations were being made to produce facsimiles. The globe gores were detached, and a tableaux (plate 19) constructed from them was incorporated in the volume when it was rebound. Although only small segments of the terrestrial globe were retrieved, included among the remnants are parts of the three gores depicting the New World on which the word "America" is inscribed on the southern continent. This represents the second use of the word on a printed map.[24]

Inspection of the three gores that focus on the New World provides significant information. On the northern continent, which is separated from the southern continent by a strait, the west coast is drawn with little detail. A line, suggesting a mountain range, extends the length of the entire continent; positioned on the line is the statement indicating that "[b]eyond the land is unknown." A geography equivalent to a Florida peninsula, Gulf Coast, and Yucatán Peninsula is recognizable. The land to the south of the gulf bears the name "Parias." The east coast is variegated, showing capes, inlets, and rivers emptying into the ocean. In the Caribbean, all of "Spagnolla" and a small portion of "Isabella" are preserved. The southern continent also includes a line denoting a western mountain range to the west of which the land is designated "terra incognita." The east coast depicts capes, inlets, and rivers. In the middle of the continent, the letters "AM ECA," from the word "America," are preserved, as are the representations of a naked child and a bird.

It is known that Johann Schöner, who was probably responsible for assembling the Wolfegg volume, produced a globe in 1515. One example is found in the Grand Ducal Library of Weimar, and another resides in the City Museum of Frankfurt. The date of 1515 is assigned because that was the year that Schöner's narrative was published. It was titled *Luculentissima quaed terrae totius descriptio . . . cum privilegio Invictis Romanorû Impera Maximiliani per acto annos: ne quis imprimat: aut imprimere procuret codices has: cum globis cosmographicis: Noriberae 1515* (A most luminous description of the whole earth . . . with the privilege of the Invincible Emperor of the Romans).

In his narrative, Schöner espouses the insular concept of the New World. The text declares: "It has now been ascertained that the earth is divided into four parts, and the first three parts are continents, that is, main lands, but the fourth part is an island because we see it surrounded on all sides by the sea." "Parias is not a part or portion of the aforesaid country, but a large independent portion of the earth, in that fourth part of the world."[25]

The Schöner 1515 globe located in Frankfurt presents a geographic depiction that is essentially identical to the segments retrieved from the binding found by Joseph Fischer. North America is named "Parias," and the South American continent bears the name "America." The two continental landmasses are separated by a sea located at about 10° north latitude. The northern continent extends from that level to 51° north latitude, and the southern continent extends to 46° south latitude. North America follows the representation of Cantino, Caveri, and Waldseemüller in offering a suggestion of a Florida peninsula and a Gulf of Mexico. Similarly, South America has a triangular form and includes an angled west coast. The land discovered by the Corte-Reals is designated as "Litus incognito" (unknown coast). Cuba is named "Isabella," and Haiti and the Dominican Republic are called "Spagnolla." Interesting features of the globe include the land named "Brasilie regio" south of the southern continent and a depiction of a strait running between that land and South America. This antedates Magellan's voyage through that strait by five years.[26]

Albrecht Dürer's Engraving of the Stabius-Heinfogel Star Chart (1515)

The Stabius-Heinfogel star chart (fig. 16), engraved by Albrecht Dürer, that Father Joseph Fischer discovered in the binding was removed from the volume and placed in the rich collection of graphics in the Wolfegg Castle. Fischer believed that the presence of the engraving by the famous artist might have been the prime stimulus for purchase of the entire volume and its contents. The complete production consists of two beautiful woodcut planispheres (projections of a sphere onto a flat surface) that together constitute the first printed star maps. One woodcut is titled "Imagines coeli Septentrionales cum duodecim imaginibus zodiaci" (Views of southern sky with twelve zodiac signs), which represents the northern half from the north ecliptic pole to the ecliptic, including the zodiac as a polar stereograph (a depiction of the brighter stars as they would be viewed by an observer). The mate, "Imagines coeli Meridionales" (Views of southern sky), depicts the south ecliptic pole to the ecliptic, minus the zodiac. Only this portion of the Dürer was present in the volume found by Joseph Fischer.[27]

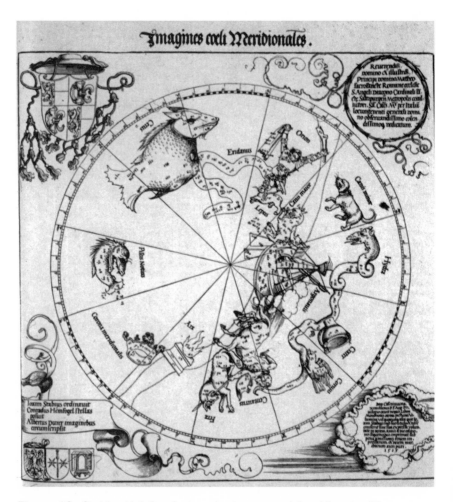

Figure 16. Stabius-Heinfogel star chart engraved by Albrecht Dürer. "Imagines coeli Meridionales," 1515, Nuremberg copperplate, 35 cm diameter on sheet 43 cm². Courtesy of Wolfegg Castle.

The work represents the collaboration of the mathematician Johann Stabius, who drew the coordinates, Conrad Heinfogel, also a mathematician, who positioned the stars, and Albrecht Dürer, who drew the constellation figures and cut the woodblocks. The maps were the first printed star maps to include coordinates and scales from which star positions could be read accurately.[28]

The "Carta Marina" of 1516

Taking up as much space as the Waldseemüller world map of 1507 in the binding at the time of its discovery in 1901 is the only known extant copy of the "Carta Marina" of 1516 (plate 20). This was also produced by Martin Waldseemüller. As is the case for the 1507 world map, the "Carta Marina" consists of twelve woodcuts on sheets measuring 45.5 centimeters × 62 centimeters (17 × 24 inches) that fit together in three zones, each containing four maps. Contained within the binding were two representations of the second sheet of the middle zone, which depicts the western portion of Africa. One was a tracing, bound in the volume, and the second was a print, loosely enclosed. The watermarks of the sheets of both large Waldseemüller maps are the same, a three-pointed crown in the shape of a fleur-de-lis. This is also true for the manuscript sheet, but the loose printed sheet of the middle zone has an anchor in a circle as a watermark. The "Carta Marina" sheets are pasted down to paper watermarked with a castle tower surmounted by a five-point coronet on a pole; above this is a second pole bearing a five-leaved fleuron.

The network of red markings is more pronounced on the "Carta Marina," which contains more additions and corrections in script. On the "Carta Marina," the margins are drawn in ink, as are the single degrees on the adjoining edges of the sheets. Elizabeth Harris's analysis (see chapters 5 and 9) leads to the conclusion that the "Carta Marina" was produced from a woodblock that did not manifest the aging of the one used for the 1507 world map, and that is why the erosion of the font used for the world map was not apparent on the "Carta Marina."

In Fischer and von Wieser's original 1903 publication, *Die Älteste Karte mit dem Namen Amerika aus dem Jahre 1507 und die Carta Marina aud dem Jahre 1516 des M. Waldseemüller (Ilacomilus)*, more space is allocated to the description of the earlier map because of its historical implication. This is in spite of the increased artistic elegance and the expanded and corrected geographical information imparted by the "Carta Marina."

Fischer and von Wieser's conclusion that Waldseemüller used the Caveri chart or its equivalent as a source for the 1507 world map is reinforced by the evidence that the "Carta Marina" incorporates information displayed on the Caveri chart. In fact, Fischer and von Wieser state: "The agreement between the Carta marina and the Canerio [Caveri] chart is so fundamental and universal, that we may directly assert: The Carta marina of Waldseemüller is a printed edition of the Canerio [Caveri] chart; not indeed a slavish reprint, but an improved and, with regard to the interior of the continents, much enlarged edition."[29]

The "Carta Marina" and the Caveri chart agree remarkably in the representation of Africa, both the northern and the southern portions, and even the dimensions of several parts. There is also agreement between the two maps related to the peninsular depiction of Greenland. For the northern part of Europe, including Jutland and the Baltic territory, it is concluded that Waldseemüller must have used a new source, possibly a Low German map, because the delineation agrees with neither the Caveri chart nor the world map. In the delineation of the Arabian Peninsula and the Persian Gulf, Waldseemüller relied on a Ptolemaic source. By contrast, the configuration of the northern part of Africa differs from Ptolemaic representations, and an even more radical change can be noted in the representation of southern Asia on the "Carta Marina." The two Indian peninsulas are presented in a more true form.[30]

In the section depicting the New World, the land outlines of the "Carta Marina" coincide with those presented on the Caveri chart. There is similarity in the representation of the land discovered by the Corte-Reals, but Waldseemüller probably used another Portuguese source in assigning the names "terra Labratoris" (land of laborer) and "terra Cortereal" (Corte-Real land). Although some of the legends from the Caveri map had been used for his 1507 map, Waldseemüller employed more of the Caveri names on the "Carta Marina" than on the earlier map. There is also remarkable congruence between the Caveri chart and the "Carta Marina" related to a crescent positioned in the north of South America.[31]

About a dozen place names have been added to the "Carta Marina" when compared with the 1507 world map, but, most poignantly, the name "America" is absent and has been replaced by "PRISILIA SIVE TERRA PAPAGALLI" (Prisilia, or the land of parrots). This is offered as evidence that the author of the map had abandoned his previous assertion, as manifest on the 1507 map, of the importance of the discoveries of Amerigo Vespucci. As previously noted, this change had already taken place on the two maps pertaining to the New World in the 1513 Ptolemy atlas. This change, regarding the importance of Vespucci, is also noted in a legend on the "Carta Marina" located near South America. It replaces the legend on the world map of 1507, on which a statement was made concerning the discovery of Brazil by Cabral (see description of sheet 10 below).

The "Carta Marina" contains a title that continues along the upper edge of the upper four sheets: "Carta marina navigatoria Portugallen, navigationes atque tocius cogniti orbis terre marisque formam naturamque situs et terminos nostris temporibus recognitos et ab antiquorum traditione differentes, eciam quor[um] vetusti non meminerunt autores, hec generaliter indicat." (A Portuguese marine navigational map, generally indicating their navigations, the lands of the known world, the form of the sea, their presently recognized positions and bounds differing from the tradition of the ancients, and also those of which ancient authors made no mention.)

Unlike Waldseemüller's world map of 1507, the "Carta Marina" is not drawn on a Ptolemaic projection but uses a rectangular network that characterized contemporary sea charts. The planispherical navigational format was not used for any printed map until 1569, when

Gerhard Mercator's large map introduced his famous projection. The presence of rhumb lines emanating from compass roses with thirty-two divisions emphasizes the sea chart aspect of the "Carta Marina," but the map should not be regarded as a simple sea chart. It is, in fact, a sophisticated world map that exceeds Waldseemüller's distinguished predecessor in both the information imparted and its artistry.

The border is an ornate presentation of shaded clouds and wind heads with dramatic and varied expressions. Fischer and von Wieser suggest that the elegance of the ornamentation bespeaks "an eminent master who unmistakably belongs to the school of Dürer." The authors proceed to point out that there are many points of similarity between the "Carta Marina" and the star map of Stabius-Heinfogel, which was engraved by Dürer. The names used for the eight main winds are Latinized versions of words that stemmed from the southern European seafaring nations. By contrast, the intermediate winds are assigned Low German terms. The artistry of the chart is amplified by regal figures, animals, and intriguing scenes. The dominant figure on the map is King Manuel of Portugal, bearing scepter and banner astride a large sea creature.[32]

Using the same numbering format that was employed in analyzing the Waldseemüller world map of 1507, each of the twelve sheets of the "Carta Marina" can be considered in a manner that permits comparison between the geography depicted on the two great maps. Most pertinent are the representations of the New World, which are shown on sheets 1, 4, 9, and 10 (on the upper left sheet of the top row, the left sheet of the second row, and the left and adjacent sheet of the bottom row). Both the eastern part of the North American continent and South America are cut off abruptly by the chart's left inner border. This obviates the opportunity of displaying the insular concept of the continental land in the New World as depicted on the 1507 world map. This is further compromised by the omission of the east coast of Asia. As a consequence, the salient and revolutionary feature of the world map of 1507—namely, two continents making up a New World interposed as a large island (on the inset) or two islands interposed between two great seas—is absent. Also, on the "Carta Marina," there is a vague broad disconnect between the northern and the southern continents instead of the narrow strait shown on the main portion of the 1507 world map and the isthmus connecting the two continents on the 1507 inset map. The absence of connection between the two continents is also at variance with Waldseemüller's 1513 "Tabula Terre Nove," on which the two continents are joined by land suggestive of Central America.

Sheet 1—extending from 280° to 337° along a longitudinal line and from "32" to "70" and climates 4–8 on a latitudinal line—is dominated by a depiction of the North American continent that bears the name "Terra de Cuba Asie Partis." This constitutes a reversion to Columbus's declaration that Cuba was a part of Asia, despite the fact that the island had been circumnavigated in 1508. The land bears the standard of the kingdom of Castile along its upper straight-line border. The east coast, depicting many named inlets and capes, abuts "Oceanus Occidentalis" and is almost identical in presentation and in the names assigned to Waldseemüller's "Tabula Terre Nove" map. The same is true for the Florida peninsula, the Gulf Coast, and the Yucatán Peninsula. This sheet of the "Carta Marina" is actually a magnification of the enigmatic geography of North America depicted on the Waldseemüller world map of 1507, that is, the presentation of a region prior to any chronicle of its discovery. On the "Carta Marina," the western portion of Cuba appears, and, as is noted on sheet 2, that island bears no name in contradistinction to the names assigned to several of the small islands in the Caribbean depicted on sheet 1.

Sheet 2 is dominated by a Ptolemaic representation of most of Europe, the Mediterranean Sea, and the north coast of Africa. The depiction of northern Europe is at variance with Ptolemy and suggests a new source. A legend in "Norbegia" (Norway) indicates that Waldseemüller planned to draw a new map of this region in the future. A strait separates England and Scotland, which bear different royal standards. The emblems of Spain, Portugal, Germany, France, Italy, Sicily, Norway, and Hungary are also shown. Southern Greenland contains a large mountain range extending the entire length, projecting down from the upper border of the map. Greenland follows the shape shown on the Caveri chart and bears the banner of Portugal. The recognition of the discoveries of Gaspar Corte-Real, sailing for Portugal, that had been made on the 1502 Cantino chart, is perpetuated on the "Carta Marina" by including a large island labeled "Terra Nova Corerati" southwest of the projecting southern portion of Greenland. The large island, part of which is shown on sheet 1, is covered with many trees; an adjacent legend specifically refers to Corte-Real's 1501 discoveries. To the south, the Azores are depicted. At the bottom of the sheet, a scale for German miles is included. In the upper right corner, a mastodonlike creature appears, and a legend indicates that a dead specimen resembling an elephant was found in that northern region. A regal figure appears south of the Barbary Coast.

Sheet 3 portrays twelve monarchs in varying poses and sizes. In the northernmost portion, scenes of a woman drawing water, a horned man with raised club chasing a hare, and a walking human with an animal's face add to the map's artistry. The geography represented on the sheet includes the eastern Mediterranean Sea, Judea, Syria, northern Egypt, Armenia, Mesopotamia, the Black Sea and the Caspian Sea, Russia with Novogard and Moscow, and a broad expanse of "Tartaria." In the left-hand margin, a scale of German and Italian miles appears. The upper right corner of sheet 4 features Tartaria, the land of the Mongols, and several encampments, one of which includes the Great Khan. Four other monarchs are depicted. Northern India is shown. An armed Mongolian warrior spurs his horse to a gallop as a dog follows. Another rider is astride a moving reindeer. There are scenes of merchants conducting dialogues, and peasants are shown participating in a variety of activities. The sheet is replete with scenes of executions by beating, fire, and decapitation. Also displayed are odd, part-human–part-animal creatures and a man whose head arises from the middle of his chest in the manner of figures in a Hieronymus Bosch painting. Several framed legends provide a description of the regions depicted.

Sheet 5 extends from 280° to 337° along a longitudinal line and from 31° north of the equator to 9° south of the equator. The geographic depiction of the continental land in the Western Hemisphere is significant in presenting a broad, ill-defined separation between the southern part of the northern continent, which is devoid of any coastal place name and the northern shore of South America. The geography of the northern coast of South America is a magnification of what was depicted on Waldseemüller's "Tabula Terre Nove." The land above the equator is named "Terra Parias" and exhibits many inlets and capes, including "cabo de las Perlas" (cape of pearls), while the land below the equator is called "Terra Nova." The latter highlights a large female opossum with a protuberant pouch and also shows scenes depicting cannibalism. That segment of land also bears the title "Terra Canibalor." In the Caribbean Sea, "Spagnolla" appears as a large named island in contradistinction to Cuba, which contains no name in spite of the fact that the smaller islands such as the Virgin Islands and "iamaiqua" are named. Reference to Christopher Columbus's discovery in 1492 appears in a legend in the region.

Sheet 6 contains a small segment of South America, including the most eastern portion of the north coast at the point where the coast assumes a southerly direction. In keeping with the Treaty of Tordesillas, only the Portuguese standard appears in the ocean to the east of the South American coast. Most of the sheet is concerned with the western portion of Africa above the equator. The coast is packed with place names designating capes, rivers, and specific locales. In the interior a rhinoceros appears along with three chiefs wearing only loincloths; also present is a fully dressed regal figure to the north in Libya. On sheet 7, latitude is marked by 31° above the equator to 10° below, and longitude extends from 34° to 93°. Parts of Libya, Ethiopia, Egypt, and Arabia are included. Unlike the orientation shown on the 1507 Waldseemüller world map, the Red Sea is depicted running almost east–west, thereby distorting the Arabian Peninsula. The east coast of Africa is displayed in detail with many place names. Three robed monarchs adorn the land farther north, while three partially clothed chieftains also appear in the south. Portuguese emblems are inserted in the waters to the east of Africa.

Sheet 8 is occupied to a large extent by India, shown in remarkable detail and including five regal feminine figures. Ceylon is located in "Sinus Gangeticus." To the east, the middle of the sheet is dominated by the upper portion of the Malaysian peninsula, which includes three monarchs and a "Sinus Magnus" (Sea of China). Most significantly, a legend states that the map was produced with imperial privilege, which extends over four years, and specifically defines 1516 ("Anno domini Millesimo quingenresimo sedecimo") as the date that the map was produced.

Sheet 9 includes no geographical elements. One half is occupied by a framed long address that identifies "MARTINVS VVALD-SEEMULER, ILACOMILVS" as the author. In the narrative, he refers to his sources and credits Columbus and Vespucci, both described as illustrious Portuguese captains, but he does not specify their discoveries. Earlier explorers, including Marco Polo, Odoric, and Piano Carpini, the last who is also associated with the Vínland map (see chapter 5), are mentioned. Most important, the text includes a direct reference to the 1507 world map and indicates that one thousand examples of the map were printed. In the legend it is stated: "Generalem igitur totius orbis typum quem ante annos paucos absolutum non sine grandi labore ex Ptolomei traditione, auctore profecto prae nimia vetustate vix nostris temporibus cognito, in lucem edideramus et in mille exemplaria exprimi curavimus . . . Additis non paucis, quae per marcum civem venetum . . . et Christoform Columbum et Americum vesputium capitaneos Portugallen lustrata fuere" (We have edited and taken care to have printed *a thousand copies of this general map of the earth, finished a few years ago* [italics mine] with great labor out of the Ptolemaic tradition by and author great in age who is hardly known in our times). It is not without a few additions, brought to light by Mark the Venetian (Marco Polo) and Christopher Columbus and Amerigo Vespucci, captains of the Portuguese.

The other half of the sheet contains two blank emblems. One

escutcheon, which is devoid of text, is surrounded by an inscription indicating that the map is dedicated to Hugo de Hassard, who was the bishop of Toul and a patron of the Gymnasium Vosagense. A second empty escutcheon of equal size appears below the first. According to Fischer and von Wieser, the lower shield is covered with a white leaf that had been pasted on. Using artificial illumination, they found that it contained a printed text, crossed out by ink, listing a series of errata.[33]

Sheet 10 includes the remainder of South America—a continuation of the portion included on sheet 7. The land is named "Brasilia sive Terra Papagalli" (Brazil or the land of parrots). The east coast is a magnification of the same segment on Waldseemüller's 1513 "Tabula Terre Nove," even ending abruptly with a straight line at about 36.5° south latitude. The flag of Castile is placed alongside the lower border. A large legend discusses the Spanish and Portuguese voyages in the 1490s, assigning priority first to Columbus, followed second by Cabral, and third by Vespucci. Toward the end of the legend, it states: "Hec per Hispanos et Portogelenses frequentatis navigationibus inuenta circa annos domini 1492 quorum capitanei fuere Cristoferus Columbus ianuensis primus. Petrus Aliares secundus. Albericusquw Vesputius tertius. Meditarranea adhuc nemo est perscrutatus . . ." (These lands were discovered by the Spanish and Portuguese in frequent voyages about the year of our Lord 1492. Their captains were the Genoese Christopher Columbus, who was the first; Petrus Aliares [Cabral] who was second, and Albericus Vespucci the third. To that time no one had ventured beyond the Mediterranean). A scale of Italian miles is included on the sheet.

Sheet 11 extends latitudinally from 10° to 50° and, longitudinally, from 34° to 93°. It is dominated by the figure of King Manuel astride a sea monster, bearing a mace and the Portuguese banner. The sheet also contains a scale of German miles, over which a large compass is spread. The depicted geography includes a detailed presentation of the southern portion of Africa. An elephant appears in essentially the same location as on the 1507 world map, and two large snakelike figures have been added. One bare-chested chieftain is seated on a throne in the eastern portion. Madagascar is inappropriately shown to the southeast of the tip of Africa.

Sheet 12 extends longitudinally from 113° to 172° and is largely occupied by one lengthy legend and seven smaller legends. The southern tip of Malaysia includes two female regal figures. In the center of the sheet is a large "GIAVA SEV LAVA INSVLA MAXIMA" (Java or a large island with lava), on which a man with cleaver in hand is disarticulating the parts of a decapitated female. The island contains the notation "Hic Antropophigox genus" (Here is a cannibal type). A large legend lists "the most notable places from which spices are transported to Calicut, the most celebrated market of all." Above the ornate border, Martin Waldseemüller is again identified as the author by the statement "Consumatum in oppido, <s< deodati compositione et digestione Martini Waldseemuller ^Ilacomili^" (Completed in the city of S. Dié composed and edited by Martin Waldseemüller, Ilacomilus).

It is quite extraordinary that Father Joseph Fischer, a savant scholar, would uncover, within a single binding, not only the holy grail of American cartography but also what some consider to be a more significant map by the same cartographer, namely, the "Carta Marina." And if that was not sufficient, the find also included segments of the second-earliest printed map with the name "America" displayed—plus the earliest star chart engraved by the greatest etcher of the time. It was not unlike uncovering the Hope diamond and, at the same time, finding, in juxtaposition, the world's largest emerald, ruby, and sapphire. With the Dürer star chart removed from the volume and placed in Wolfegg Castle's collection of graphics, the remaining cartographic gems have eventually come to rest among the treasures of the United States of America, the nation for which they played a significant role in naming.

CHAPTER 7
COMMERCE AND CONTROVERSY

What greater pleasure can there now bee, than to view those elaborate maps of Ortelius, Mercator, etc. . . . Me thinks it would please any man to look upon a Geographical map . . .

Richard Burton
Anatomy of Melancholy

COMMERCE

The discovery of a long sought after piece of history evokes excitement in the world of scholars and an equal and perhaps even more emotional response from collectors and their agents. When a find as extraordinary as that uncovered at Wolfegg Castle was made public in 1901, several societal elements reacted almost immediately. Because that find pertained to the history and heritage of the wealthiest and most powerful nation in the world, the reaction was amplified and the stakes were raised. The map responsible for the designation of continental land in the New World as "America" and considered the birth certificate for the nation that incorporates "America" in its name immediately evoked significant interest. Individual collectors and curators responsible for maintaining and expanding distinguished institutional collections pulsated with excitement at the chance to acquire the map. Book and map dealers, with antennae constantly sensing opportunities in their roles as scouts and procurers for collectors and curators, were also energized.

As to be expected, Henry N. Stevens, who at the time was responsible for acquisitions held by the most notable collectors and libraries of Americana, attempted with reflex synaptic speed to acquire the map for a client—the John Carter Brown Library—with which he had a special relationship.

Henry N. Stevens was the "Son" of Henry Stevens, Son & Stiles, the distinguished firm located at 39 Great Russell Street in London. Americana was in Henry N. Stevens's blood. His father, Henry Stevens (1819–1886), was born in Barnet, Vermont, and, according to the *Dictionary of American Biography*, was a "Bibliographer and Lover of Books." In 1845, he moved to London to facilitate his acquisition of works for collectors of Americana, including James Lenox (1800–1880) of New York City; John Carter Brown (1797–1874) of Providence, Rhode Island; George Brinley (1817–1875) of Hartford, Connecticut; and both the Smithsonian Institution and the Library of Congress in Washington, DC.[1]

During the second half of the nineteenth century, the senior Stevens acquired many treasured items for the two governmental repositories and, more particularly, for the three great collectors. Lenox's collection was eventually joined with the collection of John Jacob Astor to form the core of the New York Public Library. Henry N. Stevens continued in his father's path, seeking out items of interest for American libraries and collectors.

The correspondence between Henry N. Stevens and one of his prime clients, the John Carter Brown Library in Providence, Rhode Island, which is preserved in the archives of that library, details the sequence of the earliest attempts to acquire the Waldseemüller world map of 1507. As suggested in the first letter, dated October 26, 1901, Stevens actually learned about the discovery of the map from George Parker Winship, director of the John Carter Brown Library. Stevens wrote Winship: "My immediate interest is of course centered on your last paragraph 'What is the new Waldseemuller 1507 Mss. Map that has turned up at Wolfegg Castle?' This is the first I have heard of it. . . . This must be followed up of course."[2] A week later, Stevens wrote Winship: "By the way, he [Thacher] makes similar inquiry as you did as to the finding of a Waldseemuller map of Father Fischer. I can learn nothing in geographical circles." Thacher was a collector and historian who was noted for his pertinent book, *The Continent of America: Its Discovery and Baptism: A critical and bibliographical inquiry into the naming of America and into the growth of cosmography in New World.*[3]

In a November 7, 1901, letter from Stevens to Henry Vignaud, a French historian (who would later publish a biography of Amerigo Vespucci), it becomes evident that Stevens had traced the map. The item was of particular interest to Stevens because, ten months earlier, he had sold Mrs. S. Augusta Brown a map to be included in the John Carter Brown Library that he contended was the earliest to depict continental land in the New World—and the earliest to bear the name "America." The letter states:

> I saw a short paragraph in a London paper saying the map has been discovered at Wolfegg Castle, the residence of Prince Waldburg-Wolfegg. On looking the place up in a Gazeteer at the museum I found that Prince Waldburg-Wolfegg Waldsee does own Wolfegg Castle, which is a small place in Wertemburg. I immediately wrote off to his Highness asking for particulars of the map and for permission to see it. I shall now write Father Fischer at Feldkirch, which appears to be a small place in the Tirol about 20 miles S. of Bregenz. It is of vital importance to me to see what these maps are as they will have a great bearing on my arguments on the printed map of 1507 which I discovered as I told you. As far as I can see at present this discovery at Wolfegg seems likely to confirm my argument that the *large* Waldseemuller map described on the back of the plan in

the Cosmographie Introductio was not a printed map at all, as most
authorities seem to imagine but an advertisement of a Ms map of
which copies could be had to order by noble patrons. But we shall
see what we shall see when I see the Wolfegg map. I shall most certainly go and see it if I have the chance. I will inform you when I
get fresh news. In the meantime pray regard the matter as confiden-
tial between us, as I don't want all the world to get the start of me.[4]

Two days after writing the letter to Vignaud, Stevens wrote
George Parker Winship, indicating that, at the time, Stevens believed
that the newly discovered map was a manuscript rather than a printed
document. The letter focuses on the relationship between the map
found in Wolfegg Castle and the one Stevens had recently sold to the
John Carter Brown Library. In addition, Stevens introduces the possibility of a purchase of the newly discovered map. The letter states:

As far as the Waldseemuller map find I have followed it vigorously.
Harrisse in reply to my inquiry said he heard nothing. Vignaud sent
me a newspaper cutting from N.Y. Staats Zeitung giving the address
of Father Fischer who discovered the map at Feldkirch in the Noru-
berg Austrie. I have ascertained that the owner is Prince Waldburg-
Wolfegg-Waldsee resident at Wolfegg in Wurtemberg. I have
written to the Prince and Father Fischer asking for particulars for
permission to *go and see the map*. It is of course of great importance
to me to see these maps to see how they affect my arguments. As far
as I can see at present they confirm me exactly especially as to the
description of each plan in Cos. Intro where I suggest that the map
was never to be printed but to be a Ms one which rich patrons could
buy. If I get permission I shall take care and arrange for such bank
credit before I can go so that I can make a test whether or not the
Prince's impecuniosity or patriotism is the greater.[5]

On November 16, Stevens wrote Winship that he (Stevens) had
been granted the right to translate Fischer's work related to the map and
to publish a facsimile for distribution in England and America.[6] Continued interest in purchasing the map dominated the correspondence

from Stevens to the John Carter Brown Library over the ensuing three months. In a letter dated December 21, Stevens reports: "As to the Waldseemuller map . . . He [Fischer] says that an American has commissioned a bookseller to purchase the map from Prince Wolfegg *at any price* but as the Prince is a very rich man he will not sell them."[7]

A letter to Winship, written in January 1902, chronicles Stevens's continued efforts in this regard: "Prince would not sell except at a *very high price*. I was willing to negotiate and would show my bona fide by making a first or minimum offer of 1000 cash. . . . But what I want to know is how far the Trustees would go if the price goes beyond what I could bring myself. Of course *these are 3 things, you must have if they can be got* and I should not be surprised if the price got up to 10,000 (equivalent to today's $12.2 million), commission 10%."[8]

Later that month, Stevens wrote back, indicating that the newly discovered map was the printed document called for in the *Cosmographiae Introductio*. Reference was made to the map that had recently been purchased by the John Carter Brown Library and to Stevens's assertion that the Brown map had priority. The letter also provided an alert that others were interested in purchasing the map.

> At least 2 New Yorkers are on its track. . . . There is little doubt that the 1507 map is the veritable map described in Cos. Introd. And as such is of the greatest historical and geographical interest. . . . You will probably be wondering how this business of a new map affects your map (late mine) but you have probably come to the conclusion same as I have that our map is the earlier. I have written an article on the subject—but [there is] prejudice against the motives of anyone who has anything to sell. [I have induced] Basil H. Soulsby [to write the article], [he is] well connected, He and I are already proven personal friends for mutual interests. He would author article for Geog. Magazine for February "a friendly caveat"—neither 1st to represent in print to show the American continent nor the first to bear the name America . . . as he [Soulsby] knew absolutely nothing of the subject . . . I would inspire him. I sat up all night and having fortunately been in an inspirational mood—[9]

In that letter, Stevens indicates that he is laying down the gauntlet in defense of his assertion that the map he sold the John Carter Brown Library preceded the newly discovered Waldseemüller world map of 1507. In a series of letters written during February 1902, Stevens informed the John Carter Brown Library that the prince, who owned the map, had received numerous high offers and secured an expert evaluation that the map was worth between £10,000 and £20,000. Stevens also indicated that the prince was advised to sell at auction. Stevens believed, at that point, that the prince would be willing to negotiate if a minimum offer of £10,000 was made. Mr. Hodge of Sotheby's auction house suggested that Stevens should make a bold offer of £20,000. Stevens reflected in one letter: "Why should such things be allowed to rusticate in a German Castle when in the Brown Library they would be made accessible to students for all times?"[10]

Toward the end of that month, Stevens wrote directly to Mrs. Brown, the major benefactor of the library, indicating that he had been visited by a trustee of the library who informed him that the library could not purchase the map. The letter was obviously a direct appeal, stating: "they are the most important historical documents appertaining to America which have ever been discovered . . . there has been nothing [equivalent] since the la Cosa chart. . . . Will take commission on 10,000."[11]

On March 5, Mrs. S. Augusta Brown wrote George Parker Winship, asking him to express her regret to Stevens that she could not "become the purchaser of this Major map." Two weeks later, Stevens wrote Winship his final letter concerning the attempted purchase: "Received cable, 'declined' If I don't get it, I sincerely hope the Germans keep the darned thing. I wish they had never discovered it as it has cost me any amount of time and labor."[12]

Henry N. Stevens's status as the preeminent dealer in Americana is evidenced by the fact that his request for permission to view the map at Wolfegg Castle was honored. On September 1, 1903, Stevens and Basil Soulsby were received by Professor Fischer at Feldkirch, where they examined a facsimile of the large 1507 Waldseemüller world map. They next proceeded to Wolfegg Castle where the prince at the time displayed the volume that included the original Wald-seemüller map, which was viewed by the bibliophile-bookdealer and the map curator "with awe . . . and ever-increasing wonder."[13] The visitors then traveled to Innsbruck, where arrangements were made with the publishers of the facsimile for the firm of Henry Stevens, Son & Stiles to serve as the agent for the British and American market.

In November 1903, the twelve-sheet facsimile of the large Wald-seemüller world map of 1507 and also the twelve sheets of the 1516 "Carta Marina" by Waldseemüller were published, and each facsimile was offered for sale by the firm of Henry Stevens, Son & Stiles for £4 loose in a portfolio, or joined up and mounted on linen, folded and inserted with text in a portfolio for £5.5s (five pounds, five shillings). As stated in an advertisement:

> This work was issued in a large folio volume (21 by 15 ins.), con-
> taining facsimiles of the two maps, each in sheets (28 by 20$^{1/2}$ ins.),
> besides 2-key maps and numerous facsimiles of other ancient maps.
> The Editors' Introduction prefixed to the facsimiles fills 41 folio
> pages, followed by 12 pages of Synoptical Tables. In their learned
> Essay Professors Fischer and Von Wieser give a detailed description
> of the original maps and their discovery at Wolfegg Castle in 1901,
> followed by an account of the sources when Waldseemüller probably
> derived his information for making the two maps, and concluding
> with a long chapter on the influence exerted by them on the labours
> of subsequent cartographers.[14]

The advertisement also includes the following statement:

> Until The Discovery in 1901 of the long lost Waldseemüller Map of
> 1507 described in the *Cosmographiae Introductio* of that year, author-
> ities have always differed considerably in their conceptions as to its
> probable form and size, but no one ever suspected the existence of
> such a veritable cartographical monster as Prof. Fischer so fortu-
> nately awakened from so many centuries of peaceful slumber in the

library at Wolfegg Castle. The Map is far too large to be engraved and printed in one sheet; in fact it compromises no less than 12 sheets, each having a separate border and therefore being complete in itself. From the scope of the general design it is evident, however, that the 12 sheets were also intended to be joined together, as in a Wall Map, so as to exhibit the whole world at one glance. Each sheet measures on the average 23½ ins long by 17½ ins high at the border lines, and the complete map is four sheets long by three sheets high. When thus joined up it measures almost 8 ft long by 4 ft 6 ins high, a very large map even at the present day. The whole is engraved on wood and the quality of the work is such as to cause admiration and astonishment at the surprising development of the art at this early date. The general design, when the 12 sheets are made up as a whole, is highly pleasing and artistic. But this very fact is particularly à *propos* to the question of the priority of the Stevens-John Carter Brown Map. Even a casual glance at the facsimile will surely be sufficient to confirm Mr. Stevens's instantaneous conviction at the moment he first saw the original at Wolfegg Castle "that the St. Dié cartographers who designed the large and beautiful map could not possibly have *subsequently* drawn and prepared the Stevens-John Carter Brown Map, so comparatively crude in its design and typography. And yet there is no reason to doubt the St. Dié origin of either map.[15]

Once again, Stevens seized an opportunity to champion the priority of the John Carter Brown map. But a subsequent advertisement, which appeared in May 1907 on the occasion of the four-hundredth anniversary of the naming of America, generated greater excitement. In addition to the sale of the facsimiles for between £2 19 shillings ($14.35) in loose form to £3 14 shillings ($18) mounted on linen for exhibition as wall maps, the firm of Henry Stevens, Son & Stiles announced that it was commissioned to offer the volume containing the two original maps for sale for "Marks 1,260,000=$300,000 (equivalent to $5.6 million of today's purchasing power using Consumer Price Index) plus a 5% commission." The advertisement declares: "Owing doubtless to persistent applications, His Highness the Prince has lately recognized the fact that a permanent resting place for the Original Maps in some important American Library, where they could be more easily seen and studied, would be of greater benefit to historians, antiquarians and geographical students at large, than retaining them in the comparative seclusion of Wolfegg Castle."

Another advertisement emphasized the importance of the 1507 map in the naming of America:

> A recent author [John Boyd Thacher, *The Continent of America, Its Discovery and Baptism*, New York, 1896] figures out from the date of the *Cosmographiae Introductio*, that the New World received its baptismal name of America on the fifth day of May [April 25, new calendar] 1507, consequently the present year of Grace, 1907, is the fourth centenary of the most important incident in world's history. What more suitable or patriotic commemoration of that auspicious event could possibly be conceived than the securing of this unique contemporary memento for some representative American Library, where it could be studied by geographical students, admired by all lovers of art, and revered for all time by historians and antiquarians as the *veritable fountainhead from whence, in conjunction with the book Cosmographiae Introductio, America received its baptismal appellation!* (italics mine)
>
> All honor then to Martin Waldseemüller, who not only gave in his book its present name to the newly discovered Western Land, but also, as a geographer and cartographer, first delineated in print the outlines of that glorious discovery, and placed thereon the beautiful and time-abiding name of America, which he himself had so aptly suggested.[16]

There is no record of any offer, and it is apparent that Stevens's offer for sale was not sustained for any significant period of time. There is a report that in 1910 John Pierpont Morgan was negotiating for purchase of the map for the equivalent of fifty million dollars in today's currency, but the prince ended negotiations for the sale.[17] In 1912, Charles Heinrich, editor and general manager of Press

Exchange and the founder of the *St. Dié Press*, wrote the "Count" (Prince Wolfegg), suggesting that "Morgan, Huntington, Mellon, Carnegie, Otto Kahn, or Frank Widmer might appreciate viewing the map and purchasing it for the United States of America."[18]

The year 1912 witnessed a flurry of activity concerning the potential sale of the 1507 world map as indicated by a series of letters in the archives of Wolfegg Castle. Mr. R. Langston Douglas, a London dealer, was charged by Prince Max Waldburg-Wolfegg to sell the map, which was shipped to London where it remained for a month. At the time of transfer, Lloyds of London insured the map for £65,000. Mr. Douglas offered the map to the Library of Congress, which refused it, indicating that it did not have the resources for such a purchase. A letter from the prince to Mr. Douglas reported that there was the possibility of selling the map to Professor Bade in Berlin and that would address the concern of the German government, which "was rather shocked last time" when there was the potential for the treasure to leave Germany. The prince stated that he would not take less than 1,300,000 reichsmarks (about $200,000). Mr. Douglas wrote that the prince had been offered 800,000 reichsmarks by a Mr. Benjamin [Henry?] Stevens and that Mr. Morgan should be contacted.[19]

CONTROVERSY

Henry N. Stevens's role as an agent for procurement of the Waldseemüller world map came to an end, but in 1928, he published *The First Delineation of the New World and the First Use of the Name America on a Printed Map*, which collated and brought into focus the all-important issues of priority, pitting the map in the John Carter Brown Library against the Waldseemüller world map of 1507. The controversy concerning priority joins three other controversies of varied importance, which we have previously discussed: the controversy regarding the origin of the name, "America" (see chapter 4); the controversy about the geography depicted on the world map of 1507 (see chapter 2); and the controversy as to when the map, which the Library of Congress purchased, was actually printed (see chapter 6).

In the world of collections and collectors of printed antiquities and those involved in the sale of these items, the word *first* is of prime importance. Consequently, the establishment of priority of publication is a critical element. In the case of a document that has been purported to have "put the word 'America' on the map," priority plays a significant role in defining its value. The controversy regarding the designation of the Waldseemüller world map as the first graphic document to name America is particularly pertinent because it emanates not from a casual source or a dilettante but from a credible and distinguished bibliophile, Henry N. Stevens.

As to the issue of priority, an exhibition hall highlighting the gems of a great library's collection of printed material invariably presents as a centerpiece one of the forty-eight extant copies of the Gutenberg Bible of around 1455—if that library is fortunate to possess an example. The Gutenberg Bible represents the genesis of modern printing as the *earliest* book to use the production of movable printing type rather than individually engraved or cast letters. The book is usually opened to a page and encased in glass that filters out ultraviolet and infrared rays in order to protect the treasure. The most recently auctioned copy sold for $4.9 million in 1987 at Christie's auction house.

Similarly, special value is placed upon the Bay Psalm Book, written by Richard Mather, John Eliot, and Thomas Weld and published in Cambridge, Massachusetts, in 1640 as *The Whole Book of Psalms Faithfully Translated into English Metre*, the *first* book published in and of the thirteen colonies. The last copy to be auctioned brought $151,000 in 1971. *The Present State of New-England* by William Hubbard, published in 1677 in Boston and containing the *first* map printed in the colonies, was offered for sale in 2003 for $55,000. That copy contains the more common London edition of the map rather than the original Boston printing. The *first* printings of the Declaration of Independence, the Constitution, and the Bill of Rights are second only to their original manuscripts in significance and are among the most important documents housed in the National Archives in Washington, DC.

For the issue of priority of the Waldseemüller world map of 1507—as was the case of its discovery—chance and a prepared mind played salient roles. But, in this instance, scholarship was possibly influenced by salesmanship.

We must go back in time for a beginning. According to his own testimony, it was at a London auction in December 1893 that Henry N. Stevens made an almost incidental small acquisition. He purchased for the paltry sum of 2 pounds 4 shillings an imperfect copy of the 1513 edition of Ptolemy's *Geographia*. The item was originally thought to be of minor importance, consisting of only a small fragment of text and only one of the original twenty called-for maps. It was set aside on a shelf in Stevens's establishment where it gathered dust for six years. Then, in 1899, Stevens acquired another imperfect example of the 1513 edition of *Geographia*. This stimulated Stevens to bring forth his earlier purchase to ascertain if it could improve the more recently acquired volume.[20]

He then noticed, for the first time and with doubtless astonishment, that the map in the copy that he bought in 1893 differed significantly from the map usually found in the 1513 Ptolemy (plates 21 and 22). Stevens's example surely delineated the southern continent in the Western Hemisphere with an essentially identical outline as the usually encountered map. But within the outline of that landmass was the word "AMERICA."

At that point, Henry N. Stevens was convinced that he was in possession of the holy grail of American cartography, the long-lost Waldseemüller map, which was referred to in the *Cosmographiae Introductio*. In 1900, Stevens wrote Henry Harrisse, whom he referred to as "the greatest living authority on cartographic Americana," reporting his discovery and expounding an "interesting theory regarding his map and why it was omitted in the 1513 Ptolemy." Reference to Stevens's map appears in Henry Harrisse's personally annotated copy of his classic compendium, *The Discovery of North America*, which resides in the Library of Congress. Harrisse notes in the margin of page 444, related to the Waldseemüller map, "Besides there is another alleged Waldseemüller map of 1507, much smaller, and discovered by Stevens."[21] In 1901, that small map, which Stevens thought to be the map that named America, was offered to John Nicholas Brown, who, at the time, was in charge of Bibliotheca Americana in Providence, Rhode Island. The choice of client was fitting and natural because John Nicholas Brown's father, John Carter Brown, officially established that notable collection in 1846, in large part based on spectacular purchases emanating from the senior Stevens. The first order of Americana made by John Carter Brown consisted of some five hundred titles ranging in date from 1478 to 1794. By 1851, there had been eight such purchases, including a copy of an illustrated edition of the Columbus Letter printed in Basel in 1493, describing Christopher Columbus's initial discoveries and depicting the first illustration of the West Indies.[22]

In 1873, a year before his death, John Carter Brown, who had amassed the premier collection of books, maps, and documents relating to the discovery, exploration, and settlement of the New World, committed $150,000 for a building to house the treasures. Thus was born the John Carter Brown Library with its specific function stated in stone above the entrance: "Americana."

The year 1901 proved to be most eventful in the evolution of the controversy concerning the issue of priority of the two maps that had the name "America" placed on the southern continent in the Western Hemisphere. In January of that year, before the discovery of the Waldseemüller world map of 1507, Henry N. Stevens presented to John Nicholas Brown a 188-page report, which, to Stevens's satisfaction, proved that the map (plate 21) he was offering to the John Carter Brown Library was the first map to show any part of America, the first to name America, and was produced *before* the one that usually appears in the 1513 Ptolemy *Geographia*. Stevens also contended that the map he was offering was the lost 1507 Waldseemüller world map referred to in *Cosmographiae Introductio*. Such a bold claim and its immense importance certainly mandated endorsement by other experts. Brown, therefore, submitted the map and the documentation for assessment. The consultants concurred with Stevens's conclusions. The report was returned to Stevens for revision in anticipation of publishing the argument, but the revision was never completed, and the original report has not been found.

Before a deal could be consummated, John Nicholas Brown died. A letter from Brown's mother to George Parker Winship, director of the John Carter Brown Library, written on the stationery of the Hotel Netherland in New York City and dated February 28, 1901, says: "After all that you tell me of Mr. Stevens's research regarding the map and your conviction as to its genuineness I am ready to buy it of Mr. Stevens for £1,000 (the equivalent of £6,000 or $12,000 in today's purchasing power) feeling sure that my dear son would have done the same had he lived." The map was purchased in May 1901, and Mrs. Stevens had it preserved in an elegant morocco leather portfolio bearing the inscription "PRESENTED TO THE JOHN CARTER BROWN LIBRARY BY SOPHIA AUGUSTA BROWN IN MEMORY OF HER SON JOHN NICHOLAS BROWN 1901."[23]

In December of 1901, a momentous announcement that brought into focus the issue of cartographic primacy related to the earliest map bearing the name America appeared in print. An article written by Professor von Wieser was included in *Petermann's Mittheilungen*, titled "The oldest Map with the name America of the year 1507 and the Carta Marina of the year 1516 of Martin Waldseemüller." The article reported that the long sought after Waldseemüller map, which was announced on the title page of the 1507 *Cosmographiae Introductio*, had been discovered in 1901 by the Reverend Joseph Fischer, SJ, professor of geography at the Jesuit College in Feldkirch, Austria. Professor von Wieser wrote that the treasure was literally uncovered, by chance, while Professor Fischer was conducting research concerning the Norsemen's voyages in the library of Prince Waldburg-Wolfegg-Waldsee at Wolfegg Castle in an area of Germany that was, at the time, named Württemberg.[24]

This new discovery immediately stimulated more controversy on the issue of priority. In order to rapidly counter the claim made in the title of Professor von Wieser's article, Mr. Basil H. Soulsby, the superintendent of the map department of the British Museum, published an article, "The First Map Containing the Name America," which appeared in the February 1902 issue of the *Geographical Journal*. The publication not only served as an announcement of the prior discovery of the map that Henry N. Stevens had acquired and transmitted to the John Carter Brown Library but also expressed reservations related to Professor von Wieser's claim for the map found at the Wolfegg Castle, and suggested that an in-depth investigation should be conducted to resolve the important issue of priority. As mentioned earlier, a January 1902 letter from Stevens to the John Carter Brown curator indicates that Stevens initiated and probably wrote Soulsby's article himself. Henry Harrisse includes, in the scripted annotations of his works, a reference to the article announcing the discovery of the map in *Petermann's* and also in Soulsby's publication.[25]

In 1923, Stevens returned to the John Carter Brown Library, after an interval of more than twenty years, at which time he learned that his original analysis of the map had been lost. This occasioned his resolve to publish an essay on the map that he had sold to the John Carter Brown Library. His resolve was enforced, on his return to England, when he learned that a "Map of the World" by the Italian cartographer Giovanni Matteo Contarini, engraved by Francesco Rosselli, had been acquired by the British Museum in 1922. The names of the cartographer and engraver and the 1506 date of publication of that map appear in a Latin inscription just to the right of the Cape of Good Hope. That map was heralded as the earliest printed map to depict a part of the New World.[26]

In an article published in the *Geographical Journal* in October 1923, Edward Heawood, librarian of the Royal Geographical Society, describes the Contarini-Rosselli map and includes the following:

Note. As the priority for the representation in print of the new western discoveries is here claimed for the Contarini map, it may be

well to recall the fact that a map, probably by Waldseemüller, discovered over twenty years ago by Mr. H. N. Stevens, is thought by
him to be earlier than the big map of 1507, and to have been
printed in 1506. But as Mr. Stevens has not yet made public his
detailed arguments in support of this view it is impossible to discuss
the question of priority as between this map and Contarini's.[27]

Accepting Heawood's article as a stimulus if not a challenge,
Stevens proceeded to publish in 1928 a detailed analysis directed at
establishing the priority of the map that he had purchased in 1893
and sold to the John Carter Brown Library in 1901. In the book titled
*The First Delineation of the New World and the First Use of the Name
America on a Printed Map*, Stevens addresses six basic hypotheses: (1)
The Stevens-John Carter Brown map was printed before the map
titled "Orbis Typus Iuxta Hydrographorum Traditionem," which
appears in all other extant copies of the 1513 Ptolemy *Geographia*; (2)
sometime between 1505 and 1506, Martin Waldseemüller and his
associates at the Gymnasium Vosagense constructed the Stevens-John
Carter Brown map before the establishment of the printing facility at
the gymnasium and, therefore, before that press's initial production,
Cosmographiae Introductio, which was completed in April 1507; (3)
the Stevens-John Carter Brown map was printed in Nuremberg using
Nuremberg type on Nuremberg paper; (4) the Stevens-John Carter
Brown map represented an experimental design for maps to be incorporated in a proposed new edition of Ptolemy's *Geographia*. Proof of
these four hypotheses would justify the summary declared as the fifth
and sixth assertions, namely, that the Stevens-John Carter Brown
map merits the recognition as the first known printed map to show
any part of the New World as well as the first known printed map on
which the name "America" appears.[28]

     Although the Stevens-John Carter Brown map was initially pur-
chased as part of an incomplete 1513 Ptolemy *Geographia*, Stevens
contends that the map was clearly an insertion as evidenced by the
fact that it is printed on paper with a watermark that differs from the

watermark on the papers that make up the rest of the volume. Also, the paper on which the map is imprinted is almost a half inch smaller than other pages, and the map page contains wormholes, which are absent from the end binding and from the remainder of the volume.

When the paper of the Stevens-John Carter Brown map is compared with that of the map without the name "America" usually found in the 1513 Ptolemy, a distinct difference is apparent. The paper on which the Stevens-John Carter Brown map is imprinted contains a large watermark consisting of a large castle tower surmounted by a five-point coronet on a pole, above which is a second pole bearing a five-leaved fleuron. The watermark of the usually encountered 1513 "Orbis Typus Universalis Iuxta Hydrographorum Traditionem," by contrast, is much smaller and is simply a coronet with three points having the form of a fleur-de-lis.[29]

Additional support for the hypothesis that the Stevens-John Carter Brown variant antedates the usually encountered 1513 map is found in Charles Moïse Briquet's tome on watermarks, *Les Filigranes, Dictionnaire historique des Marques de Papier* (Paris 1907). That book, which includes an in-depth analysis of paper produced in the fifteenth and sixteenth century, provides evidence that the paper used for the Stevens-John Carter Brown map was employed from 1477 to 1511. The watermark present in the paper on which the Stevens-John Carter Brown was printed is also encountered on several pages in the famous Latin Bible printed by Anthony Coburger in Nuremberg in 1477 and on Albrecht Dürer's three great books of 1511, the *Great Passion, Life of the Virgin*, and *Apocalypse*, also printed in Nuremberg.[30]

Stevens, however, was unaware that the twelve sheets of the "Carta Marina" of 1516 bearing the three-pointed crown were pasted onto twelve sheets of paper bearing watermarks of a castle tower surmounted by a five-point coronet on a pole. Above that crown is a second pole bearing a five-leaved fleuron, which is identical to that present on the paper used for the Stevens-John Carter Brown map. This would suggest that the map in the John Carter Brown Library was produced in 1516 or later.

224 PUTTING "AMERICA" ON THE MAP

At first glance, the Stevens-John Carter Brown and the usually encountered 1513 map present deceptive similarities. The two maps share the same title appearing in identical type. The place names, "OCEANUS OCCIDENTALIS," "MARE INDICUM," "GRON-LAND," " EVROPA," "ASIA," and "AFRICA" are written in the same large capital letters, while *"Tropicus," "Capricorni,"* and *"Equinoctialis Circulus"* are identical. Henry N. Stevens, however, presents evidence of sufficient discrepancies to conclude that the two maps were printed from different woodblocks using distinctly different techniques.

In the case of the usual 1513 map, all the place names are imprints of Gothic letters, which were directly engraved on the woodblock. By contrast, the names on the Stevens-John Carter Brown map were printed from roman type that was inserted on pegs into holes piercing the block. The difference in technique is made apparent by an analysis of the relationship between the names and the loxodromic lines. On the usual 1513 map, the loxodromic lines (lines that sailing vessels take; the lines run oblique to the equator and cross all the meridians at the same angles), which were engraved on the wood-block at the same time, pass through the names without a break. By contrast, on the Stevens-John Carter Brown map, a small space characterizes each intersection between a name and a line. The spaces are a consequence of the lines having been cut away to allow the type to pass through them.

When compared with the Stevens-John Carter Brown map, there are many inaccuracies in the printed type on the usual 1513 Ptolemy map thought to be the result of the difference in techniques employed for the two maps. In the case of the Stevens-John Carter Brown map, a pierced block system was used in which the type was set from an actual drawing of the map or a proof of the outline block with the letters written with the correct orientation. Whereas in the case of the usually encountered Ptolemy map, the letters had to be carved directly on the woodblock in reverse. Errors in letters engraved directly on the block would be more difficult to correct.

Analysis of the lettering on the Stevens-John Carter Brown map leads to the conclusion that all the place names use the same lettering, and that the name "AMERICA" was inserted at the same time as the other place names with the identical capital font. This suggests that all the lettering took place at the same time. The critical question, however, is, Was the Stevens-John Carter Brown map an early prototype of the map that appears in all other copies of the 1513 Ptolemy or does it represent a later correction that incorporated the name "America"?

In defense of his thesis that the Stevens-John Carter Brown map is prototypical, Henry N. Stevens, in addition to evidence that the map was printed on paper that was used in Nuremberg prior to 1513, points out that the type employed for the names on that map was the same as that used for several books printed at Nuremberg between 1501 and 1510. Stevens could not find the specific roman type used on the map in the John Carter Brown Library in any book in the British Museum printed in Nuremberg after 1510. The type is distinct from that used at Saint Dié or Strasbourg, suggesting that the Stevens-John Carter Brown map was printed at Nuremberg before the introduction of printing in Saint Dié and before *Cosmographiae Introductio* came off the Saint Dié press, as its first production, in April 1507.

In addition, Stevens uses a comparison of the place names that appear on the two maps in making the case that the map currently in the John Carter Brown Library antedates the similar map that appears in all other extant copies of the 1513 Ptolemy *Geographia*. Of the 160 names found collectively on the two maps, there are two on the Stevens-John Carter Brown map that are absent from the usual 1513 Ptolemy map; there are eight that are present on the latter and missing on the former, and there are five names that are entirely different.

The two names exclusive to the Stevens-John Carter Brown map are "America" and "Venetia." Stevens contends that if the map had been produced at Saint Dié between 1505 and 1506, as he concludes, it would have been natural for the word "America" to appear, because, at the time, the members of the Gymnasium Vosagense had assimilated material on the Vespuccian voyages. By contrast, the map that appears in all other copies of the 1513 Ptolemy atlas was produced at Strasbourg, and was printed at a time that Waldseemüller, the cartographer, did not include Vespucci's name on his other map of the Western Hemisphere, "Tabula Terre Nove," in that atlas. In fact, on the map titled "Tabula Terre Nove," the cartographer specifically credits Christopher Columbus with having discovered the continent.

The fact that "Venetia" is present on the Stevens-John Carter Brown map and absent from the 1513 Ptolemy is offered by Stevens as more proof of the priority of the former. Because no other cities in Europe are designated on the Stevens-John Carter Brown map, it is assumed that "Venetia" refers to the important province of the Republic of Venice. Between 1508 and 1513, the Republic of Venice was partitioned by Emperor Maximilian and other European rulers and reduced to the city of Venice. Consequently, it was omitted from the usual 1513 Ptolemy "Orbis typus" map, which does not name cities. "Venetia" is similarly omitted from the modern large-scale map of Italy, which Waldseemüller produced for the same 1513 Ptolemy atlas.

There are eight names present on the map usually appearing in the 1513 atlas and absent from the Stevens-John Carter Brown map. Three of those names, "*Barbaria*" (the African town), "*Rio de Lago*" (southeast Africa), and "*Cananor*" (on the west side of the Indian peninsula), also appear on other maps in the 1513 Ptolemy atlas. When the five names that differ on the two maps are compared for priority, Stevens notes that those on the map usually found in the 1513 atlas are more modern. "*Cefare*" and "*Cirene*," which are early Ptolemaic names of two North African provinces appearing on the Stevens-John Carter Brown map, are replaced on the 1513 Ptolemy map by "*Sera*" and "*Zanara*"—the more modern names of the two towns. Similarly, "*Fortunatæ*," "*Littouia*," and "*perithos gallas*" are replaced by the more modern "*Canaree*," "*litluania*," and "*paticho S galla.*"

Reinforcing the argument that the usually encountered 1513 map was constructed later than the Stevens-John Carter Brown map is the fact that the block that was used for the usually encountered 1513 map continued to be employed for the 1520 edition of the Ptolemy *Geographia*.

Continuing the defense of his thesis regarding the priority of the map in the John Carter Brown Library, Stevens provides evidence that the participants at the Gymnasium Vosagense in Saint Dié had initiated their work on a proposed new edition of Ptolemy before 1507. In a letter written to Johann Amerbach of Basel on April 7, 1507, *antedating* the publication of *Cosmographiae Introductio*, Waldseemüller wrote: "I think you know already that I am on the point to print in the town of St. Die the Cosmography of Ptolemy, after having added some new maps." Also, early in 1507, a book titled *Speculi Orbis . . . Declaratio* by Gaultier Lud, canon of Saint Dié, was published in Strasbourg. That work states: "1. that a figure of the unknown country recently discovered by the King of Portugal has been hurriedly prepared; 2. that a more detailed and exact representation of that coast would be seen in the new edition of Ptolemy; 3. that the new edition of Ptolemy would soon be prepared."[31]

Although, the "hurriedly prepared" map referred to in the *Speculi Orbis* cannot be precisely identified, Stevens contends that the statement suggests that some map had been prepared by early 1507 and that a more detailed map would soon appear in a new edition of Ptolemy. Stevens indicates that it is highly unlikely that the map referred to as "hurriedly prepared" was the elegant twelve-sheet 1507 Waldseemüller world map.

In his effort to establish the fact the Stevens-John Carter Brown map not only antedates the 1513 Ptolemy map but also was produced before *Cosmographiae Introductio* came off the press at Saint Dié, Henry N. Stevens dissects the text contained within the 1513 Ptolemy *Geographia*. In their ad lectorem (address to the reader), the editors of the 1513 Ptolemy, Jacobus Eszler and Georgius Übelin, indicate that the work was in preparation before 1507, at which time it was suspended. Mention is also made that a prototype of the "Orbis Typus Universalis Iuxta" was "given out to be engraved for the press . . . through the generous assistance, whilst he lived, of René most illustrious Duke of Lorraine, now piously deceased." The duke died in 1508. Stevens argues that this indicates that a prototypical map was produced before that date.

Comparative analysis with the maps included in the 1513 Ptolemy reveals that the technique used for the Stevens-John Carter Brown map was abandoned for production of the atlas. Stevens deduces that the insertion of type letterings, used for the Stevens-John Carter Brown map, was found to be impractical when a large number of names had to be incorporated in the 1513 atlas. He concludes that when the 1513 edition of Ptolemy was printed at Strasbourg, the so-called "Orbis Typus" map was recut, perhaps because the original block that was made before 1507 was weakened by the insertions. When the new map was prepared, the word "America" was removed because during the ensuing six years, the cartographer, Waldseemüller, appreciated that it was a misnomer. This is substantiated by the absence of "America" on the "Tabula Terre Nove" appearing in the same 1513 Ptolemy.

Thus, the comparison of techniques; the identification of the type and watermark of the paper used for the Stevens-John Carter Brown map indicating that the map was printed at Nuremberg before 1510; the comparison of place names on that map with those appearing on the 1513 Ptolemy map; the historical events that took place at the gymnasium in Saint Dié; and the interpretation of the text and analysis of the maps of the 1513 Ptolemy are offered as evidence substantiating the conclusion of Henry N. Stevens that the Stevens-John Carter Brown map was engraved in 1505–1506, before the publication of *Cosmographiae Introductio*.

With the firm conviction that the map in the John Carter Brown Library antedates by at least seven years the map that is usually included in the 1513 Ptolemy *Geographia*, Stevens next set out to address the issue of priority between the map in the John Carter Brown Library and the large 1507 Waldseemüller world map, which at the time resided in Wolfegg Castle. Both the 1507 map and the map in the John Carter Brown Library are, with almost complete certainty, the works of Martin Waldseemüller. The critical questions to be resolved are, which of the two preceded the other in depicting a part of the New World in print, and which document earns the designation as the first printed map to assign the name "America" to the newly discovered land?

It is generally thought that, because of its complexity, the large 1507 Waldseemüller world map was printed on twelve sheets by Johann Grüninger at Strasbourg rather than at Saint Dié. Assuming that to be true, Stevens asks the rhetorical question: "If the map in the John Carter Brown Library was engraved and printed at Nuremberg (as he had previously concluded), after the introduction of printing into St. Dié in the spring of 1507, why did the St. Dié cartographers go so far afield (some two hundred miles) to have their crude pattern-block made for the Ptolemy, when they already had the experience of engraving and printing (in a much more artistic and up-to-date manner, either at home or at Strasbourg, less than fifty miles away) the twelve sheets of the large map?"[32]

The large Waldseemüller world map is unquestionably the more sophisticated map. It is the earliest printed map that attempts to include the entire 360 degrees of longitude on its marginal scale. Using a modification of the coniform or cordiform projection of Ptolemy, the longitudinal meridian lines come together at the North Pole. By contrast, the Stevens-John Carter Brown map is drawn on a rectangular projection that had been used on early manuscript marine charts, known as portolans.

The large wall map actually presents two geographical representations of the New World. On the small inset map in the upper border depicting the Western Hemisphere, or the Vespuccian world, the northern and southern continents are joined by an isthmus; an irregular line cuts off the lower end of the hemisphere; and the name "America" does not appear on the southern continent. By contrast, on the main map, the two continents are separated by a body of water and the name "America" boldly stands out on the southern continent, which extends farther south than on the inset.

The depiction of geographic elements in the New World is certainly more inclusive on the large Waldseemüller world map when compared to the Stevens-John Carter Brown map. On the latter, the southern continent contains only four coastal place names and is abruptly cut off with a straight horizontal line at about 371/2° south latitude, while the sea extends to 80° south latitude. Four named islands—"*Isabella*," "*spagnolla,* " "*Gigant*," and "*Brasil*"—appear north of the equator, and the Tropic of Cancer runs just south of the island of "*spagnolla*." In the mid-Atlantic, to the west of Greenland, there is a small piece of land with only the east coast defined, which might represent part of the North American continent.

On the Waldseemüller world map of 1507, the Tropic of Cancer runs through the island of "*Isabella*." Many islands in the Caribbean Sea are named. Above the strait separating a northern from a southern continent, the word "*PARIAS*" appears in the region of Venezuela, where, according to the misinformation contained in the *Lettera of Vespucci*, Vespucci allegedly landed in 1497, a year before Columbus. There is what might be a representation of a Gulf of Mexico and a Florida peninsula, and multiple names are imprinted along the east coast of the northern continent. Even more names appear along the coast of the southern continent, which extends farther south than on the Stevens-John Carter Brown map.

Both the northern and the southern continents in the Western Hemisphere on the Waldseemüller map are interposed between two oceans. Along the west coast of the two continents, the words "Terra ultra incognita" (unknown land) appear, and, in addition to "AMERICA" on the southern continent, the statement "TOTA ISTA PROVINCIA INVENTA EST PER MANDATUM REGIS CASTELLE" (All of this province was discovered by mandate of the King of Castile) is included.

With the priority of the Stevens-John Carter Brown map in relation to both the 1513 Ptolemy "Orbis typus" map and the 1507 Waldseemüller world map established to his satisfaction, Henry N. Stevens argues that the Contarini-Rosselli map—bearing the date 1506 and purchased by the British Museum in 1922—was also produced subsequent to the Stevens-John Carter Brown map.[33]

The precise date of the Contarini-Rosselli map is difficult to define because of disparities in the calendar in various locations in Europe early in the sixteenth century. In the case of the Waldseemüller map, if it is assumed that it was issued with the *Cosmographiae Introductio* at the time of the first edition—which was dated VII Kalend Maii (April 25) 1507—we have a specific date of reference. By contrast, the Contarini-Rosselli map contains only the year 1506 with no month. The Contarini-Rosselli map was probably printed in Florence or Venice. If the map was produced in Florence, where the paper was made and where the engraver, Rosselli, was located, the printed date of 1506 could have referred to any date between March 25, 1506, and March 24, 1507, in the new calendar. Therefore, it is not even certain that the Contarini-Rosselli map antedates the Waldseemüller map by a whole year.

Invoking the issue of technique, Stevens proposes that the Waldseemüller world map of 1507 might have been initiated before the Contarini-Rosselli map. The former consists of twelve sheets and was produced from an engraving on wood, a much slower process than that of copperplate engraving, which was used for the latter one-sheet map. It is conceivable that the two maps were begun at about the same time, but the Contarini-Rosselli map, requiring less time to design, engrave, and print, would have appeared earlier. Henry N. Stevens contends: "It has already been shown that the Stevens-John Carter Brown Map is earlier than the large Waldseemüller, hence unless it can be proved that the Contarini map is earlier than the large one not only in its publication but in its *conception* we have *prima facie* a logical deduction that the Contarini cannot be as early as the Stevens-John Carter Brown map."[34]

Also, the Contarini-Rosselli map is presented as an extension of the 180 degrees of the old Ptolemaic projection while the Stevens-John Carter Brown is drawn on a more modern rectangular projection similar to manuscript sea charts of the period. In addition, the configurations and nomenclature on the Contarini-Rosselli are more refined, thereby supporting the priority of the Stevens-John Carter Brown map.

Thus, Henry N. Stevens had made, to his own satisfaction, a

strong case for his multifaceted thesis that the map, which he sold to the John Carter Brown Library in 1901, was the prototype of the "Orbis typus" map that appears in all other copies of the 1513 Ptolemy *Geographia*. Stevens deduced that the map residing in the John Carter Brown Library was designed and prepared by Martin Waldseemüller at Saint Dié sometime between 1505 and 1506 and engraved and printed at Nuremberg no later than early 1506. As such, it merits designation as the earliest known printed map to depict part of the New World and the first map on which the word "America" appears.

Although the scholarship is in-depth and impressive, the fact remains that it was conducted by an individual with a vested interest. Henry N. Stevens, the scholar, was also the salesman, who sold the map to the John Carter Brown Library, doubtless at a price commensurate with its importance. The expected result is that two disparate groups of critics have surfaced: one in agreement with Stevens's conclusion and the other expressing strong disagreement.

Two months after the discovery of the 1507 Waldseemüller map was announced in *Petermann's Mittheilungen*, an article by Basil Soulsby was published in the *Geographical Journal*. Soulsby was assistant in the map room at the British Museum and honorary secretary to the Hakluyt Society. Predictably, Soulsby strongly endorsed the conclusion that Stevens had made at the time of the initial viewing of the Waldseemüller map in 1903, a conclusion that led to the research and writings that culminated in Stevens's 1923 book. In his 1902 article, Soulsby devoted the first four pages to a description of the Waldseemüller map, to its genesis and discovery. The remaining five pages are a direct endorsement of the conclusions articulated by Stevens. The article challenged the claim that the newly discovered world map offered the first representation in print of New World discoveries and that it was the first map to name "America." Soulsby, in essence, drew from an essay that Stevens had given to G. F. Barwick, assistant keeper of printed books in the British Museum, and according to Stevens's letter to Winship, Stevens played a large role in the development of the article.

Sharp criticism of Stevens's scholarship and conclusions came from Louis C. Karpinski, a respected cartographic historian at the William L. Clements Library in Ann Arbor, Michigan. In addressing Stevens's evidence that the paper and the type of the map in the John Carter Brown Library was used only between 1505 and 1506, Karpinski reports that he found that the same paper and type was used for a copy of Apianus's *Astronomicum Caesareum*, published in Ingolstadt in 1540. This discovery and the suggestion that names (which were generally of minor importance and appeared on the usually encountered 1513 Ptolemy map) were omitted on the Stevens-John Carter Brown map to save work on the woodblock argue against the conclusion that the Stevens-John Carter Brown map was the prototype for the 1513 map.

Karpinski writes: "The likelihood that Waldseemüller put the name America upon the New World in advance of the publication suggesting the name America is remote. It could almost be qualified as absurd. The likelihood that he and his colleagues prepared a woodblock with type inserts as a specimen block exists as a historical fact only in the mind of Mr. Stevens."[35]

In his argument, Stevens invokes the writing of Matthias Ringmann. Ringmann, in his letter dated August 1, 1505, to Jacob Braun, which appeared in the Vespuccian tract *De Ora Anarctica*, stated: "This little book of Albericus [i.e., Vespucius] fallen by chance in our way we have read through hastily and have compared almost all items with Ptolemy (whose maps as you know we are now examining with diligence)." Stevens offers this as evidence that the map in the John Carter Brown Library was in preparation at the time, and that Vespucci received credit for discovery in the New World.[36]

Countering this conclusion, Karpinski points out that Ringmann did not go to Saint Dié until the end of 1506 or early 1507 and that the reference merely indicates the beginning of a study. Also "Albericus" and not the word "America," as it appears on the Stevens-John Carter Brown map, was used as the Christian name of Vespucci at the time. Karpinski also indicates that Waldseemüller, in his dedication to Emperor Maximilian in the *Cosmographiae Introductio*, makes reference to "a figure of the entire earth as well in the form of a globe as representation on a plane." No mention is made of any other map by that cartographer pertaining to the New World. Similarly, no mention is made of a world map in Waldseemüller's 1508 letter to Ringmann that appeared in the 1508 edition of *Margarita Philosophica* by Gregorius Reisch, other than the two maps accompanying the *Cosmographiae Introductio*.

Karpinski also draws from the differences in the geographical configurations of the 1507 Waldseemüller world map and the Stevens-John Carter Brown map. In 1510 and 1512, respectively, Glareanus, on a manuscript, and Stobnicza, on a printed map, replicated the depiction of the 1507 Waldseemüller world map. On the other hand, no map produced before 1513 that follows the format of the Stevens-John Carter Brown map has been uncovered. The map usually appearing in the 1513 Ptolemy *Geographia* and the Stevens-John Carter Brown map, according to Karpinski, represents an intermediate step between the 1507 Waldseemüller world map and the "Carta Marina" that Waldseemüller produced in 1516.

On the 1507 Waldseemüller map, according to Karpinski: "Scotland bears a curious protruding extension to the East; Italy has a strange curve downwards at the heel; the Indian peninsula is practically nonexistent; the island of [Ceylon] Taprobana is enormous; the Red Sea runs almost straight north and south; the northern coast of Africa slopes from northwest to southeast." Each of these depictions can be noted in the 1490 edition of Ptolemy and are corrected on the "Carta Marina" of 1516. The errors are also corrected, at times, only partially on the Stevens-John Carter Brown map. This constitutes evidence of its status as an intermediate step subsequent to the Waldseemüller world map. Karpinski opines, "It is inconceivable that a scholar of the ability of Waldseemüller should have made in 1506 a map as correct as the Stevens-John Carter Brown map and then in 1507 have gone back to the absurdities of the old Ptolemaic geography."[37]

Championing the singular importance of the 1507 Waldseemüller world map, Karpinski concludes:

There is no intention in this article of casting any discredit upon the account given by Mr. Stevens. However, in historical matters, documents that are as definitely dated as the Castle Wolfegg Wald-seemüller map of 1507 and the Contarini map of 1506 at the British Museum are rightly accorded an authority that cannot be given to some document whose dating involves elaborately spun theories. The accidental placing of the name America by some unknown printer on a map current from 1513 to 1541 has little significance either in the history of American cartography or the history of the naming of America. The majestic map issued with certainty in 1507 began a new period in cartography of the New World and placed the name America definitely upon the map of the world.[38]

The late R. A. Skelton, superintendent of the map room of the British Museum, offers the most recent assessment of the issue of the priority of the Stevens-John Carter Brown map when compared with the 1507 Waldseemüller world map in the 1966 publication of the facsimile edition of *Claudius Ptolemaeus Geographia Strassburg*. Skelton accepts the fact that Stevens proved that the map in the John Carter Brown Library was engraved and printed in Nuremberg, prob- ably as a trial block, no later than 1507. Skelton, however, regards Stevens's attempt to associate the Stevens-John Carter Brown map with the "hastily prepared representation" referred to by Lud in 1507 and also Stevens's argument that Duke René paid for the engraving of that map and other modern maps in the 1513 Ptolemy as uncon-vincing. Skelton also considers it doubtful that an analysis of the technical aspects of the two maps helps determine their chronology.

Which document merits designation as the Birth Certificate or Baptismal Document of the United States of America? A birth cer-tificate mandates the inclusion of two specific items, a name and a date, which throughout the life of an individual remain as points of reference and identification. The great Waldseemüller world map pro-vides the name "America" and a date, based on the reference to the map in *Cosmographiae Introductio*, of April 25, 1507, both of which remain uncontested. The Contarini-Rosselli map does bear the date

1506, but no part of the current United States or South America is depicted, and the name "America" is absent.

The Stevens-John Carter Brown map includes the name "America" on a partially depicted southern continent in the Western Hemisphere but shows only an equivocal continental landmass in the northern half of that hemisphere, and it is associated with only a deduced and contested date. The large 1507 Waldseemüller world map unequivocally depicts two continents in the Western Hemisphere interposed between two great bodies of water. Each continent has an Atlantic coast; each has a west coast, surprisingly depicted before Balboa's sighting of the Pacific Ocean from a peak on the Darien Peninsula (Panama) in 1513. The continents are separated by a strait on the main map but are correctly joined by an isthmus on the inset devoted to the Western Hemisphere. A Florida peninsula and the Gulf Coast can be inferred by inspecting that map. Any American schoolchild or an unsophisticated adult viewing the Waldseemüller map would readily appreciate that he is looking at his country, albeit in a somewhat primitive and distorted form.

Baptism is, by definition, a Christian sacrament signifying spiritual rebirth. Beginning with the discoveries of Christopher Columbus in 1492, the lands of the Western Hemisphere that had been the domain of the Native Americans for almost ten millennia were claimed and controlled by the Christian European powers. The baptismal document for what Europe regarded as a New World merits a fittingly elegant and striking presentation, which provides both a Christian name—honoring a Christian explorer—and a recognized date. Only the large 1507 Waldseemüller world map fulfills those criteria. Even the staunchest of antagonists, Henry N. Stevens, in his 1907 advertisement for a proposed sale of the Waldseemüller map, refers to the map as "the veritable fountainhead from whence, in conjunction with the book, *Cosmographiae Introductio*, America received its baptismal appellation!" Is it not fortunate and a source of pride that the birth certificate and documentation of the baptism of the land on which the United States of America evolved is a glorious map, a grand piece of art measuring approximately thirty-four square feet? It is a document that was viewed at the Smithsonian Institution by hundreds of thousands of citizens on the five-hundredth anniversary of Columbus's discovery in 1992. It will forever remain available for viewing at the Library of Congress for current and future generations of citizens and visitors to this country. Truly, the greatest of modern nations has a most impressive pictorial certificate of its birth.

CHAPTER 8
NUDGING AND NEGOTIATIONS

As geographers, Sosius, crowd into the edges
Of their maps of the world which they do not
know about, adding notes in the margins to the
effect this lies nothing but sand deserts, full of
Wild beasts and unapproachable bogs.

Plutarch's Lives

Despite Henry N. Stevens's book, the controversy over priority of the first printed map to depict land in the New World and the first to put the name "America" on a map did not extinguish the desire to acquire the Waldseemüller world map of 1507. In the papers of the Clements Library in Ann Arbor, Michigan, there is a letter written in 1931 from a Mr. Peitz of Germany to Louis C. Karpinski, who worked at the library. It states that the duke (Prince Waldburg-Wolfegg) is not opposed to selling the atlas to Americans, "but it must be handled discreetly."[1]

According to *Rosenbach: A Biography*, in the spring of 1933, "An agent for Count Moy, who was agent for the Prince of Waldburg-Wolfegg, made overtures to sell the incomparable wall-map of 1507 on

which the name America appears for the first time. Once it would have
sent the Doctor [Rosenbach] flying out to see Mr. [Henry E.] Huntington,
(a notable collector for whom he had acted as an agent) in a flurry of
excitable hundreds of thousands. Now the overture remained unanswered."[2] In 1938, Erwin Raisz, lecturer in cartography at Harvard's Insti-
tute of Geographical Exploration, wrote inquiring about the possibility of
purchase.[3] Willis Van de Vanter, curator of Paul Mellon's collection,
indicated that Mr. Mellon actually tried to buy the Waldseemüller map
in the 1960s, but the count was not interested is selling at the time.[4] (At
this time a count directed the family's fortunes and activities.)

The archival material in the Library of Congress, which chronicles the library's association with the volume containing the Waldseemüller world map of 1507 and its companions, from the time of
discovery of the map to its ultimate purchase and transfer, extends
more than one hundred years. In the annual report of June 30, 1904,
Philip Lee Phillips, the first chief of the Geography and Map Division,
states in an internal memo: "In the summer of 1901 Professor Joseph
Fischer discovered at Wolfegg Castle, Austria [sic] copies of Waldseemüller's maps of 1507 and 1516 thus ending the most famous cartographical search of the present age."[5] There is evidence of the map
having been offered to the Library of Congress in 1912 (see chapter
7). Also, there is a note indicating that sometime between 1928 and
1938, the library made an offer for the map, but it was rejected.[6]

Immediately after World War II, Burton Atkinson, chief of the
Geography and Map Division, anxiously asked the State Department
to determine the status of the 1507 world map. He was relieved to
learn that it was safely preserved in the castle. In 1976–1977 the
Librarian of Congress, Daniel Boorstin, and Walter Ristow, then chief
of the Geography and Map Division, explored the possibility of purchasing the map. This probing included Walter Ristow's visit to
Wolfegg Castle. At that time, a potential offer of one million dollars
was discussed, but the map's owner, Count Max Willibald Waldburg-
Wolfegg, was adamantly opposed to any sale.[7]

It is speculated that the year 1983 witnessed a relaxation in the count's attitude toward a potential sale of the map to the Library of Congress. With the personal encouragement of Ambassador William H. G. Fitzgerald, Count Max Willibald Waldburg-Wolfegg generously allowed the volume containing Martin Waldseemüller's world map of 1507 and the 1516 "Carta Marina" to leave Wolfegg Castle and travel to the Western Hemisphere for the first time since its acquisition in order to highlight an exhibit titled "The Naming of America." With the exception of the one occasion when it was shipped to London because of the potential for a sale, the volume, including the Wald-seemüller map, had spent the first four centuries of its existence locked in a safe in Wolfegg Castle, and the map had never been assembled by joining its twelve sheets.

"The Naming of America" exhibit was displayed at the National Museum of American History, Smithsonian Institution from October 1983 to February 1986, and consisted of an incomparable array of items related to the discovery and mapping of America. The title panel of the exhibit emphasized the importance of the map.

In 1500, the newly discovered lands of the New World lacked a name. They were known by various labels: Parias, Cuba, Brasilia. It was in 1507 that the name "America" was first put forward in a book, on a world map, and on the paper segments for a globe, all produced in St. Dié, France. Martin Waldseemüller, a German scholar, is given credit for drawing the map and globe, both of which were lost until recent times.

In 1901, at the 16th century Wolfegg Castle in southern Germany, a volume was found containing a sea chart dated 1516 and a map that is thought to be the long-lost world map of 1507. But the map is unsigned and undated, so until all the evidence is in, questions remain open: who made this map and when, and is it the St. Dié map of 1507 or perhaps a later edition of it?

The map is now shown publicly for the first time.

The first of the exhibit's four sections included a scale model of Christopher Columbus's flagship, the *Santa Maria*, and a copy of the

1495 Florentine edition of Columbus's letter to his sovereigns. It was opened to a woodcut portraying the king as the patron for the expedition, the explorer's fleet, the discovered islands and their native inhabitants. A 1508 Basel edition of *Mundus Novus* completed the section.

The second section bore the title "Mapping the Known World." It included a 1482 Ulm edition of Ptolemy's *Geographia* and the manuscript world map of Martellus (circa 1489) representing the transition between Ptolemaic geography and the works of cartographers, who depicted the New World on their maps. Also displayed was a replica of the 1492 globe by Martin Behaim presenting the world on the eve of Columbus's discoveries. It had been on display earlier in the World's Columbian Exposition held in Chicago in 1893.

Also appearing in that section were the Lenox globe and facsimiles of the manuscript Caveri 1504–1505 map, Bartholomew Columbus's sketch of the New World (circa 1506), and the Contarini-Rosselli world map of 1506. Matthias Ringmann's translation of a Vespucci letter, published as *De Ora Antarctica* (see chapter 1) was included as another example of *Mundus Novus*.

The third and salient section included a narrative about the group of scholars who had assembled at the Gymnasium Vosagense in St. Dié and had a copy of the *Cosmographiae Introductio* on display. The centerpiece was a facsimile of the Waldseemüller world map of 1507 with the twelve sheets joined to present the entire map as a unit. It was accompanied by one of the original Waldseemüller globe gores of 1507. The manuscript map of Glareanus (circa 1510), the two hemispheres of Stobnicza (1512), and the 1520 map by Peter Apian[us] were all on display, as was the facsimile of the 1516 "Carta Marina."

Johann Schöner, who is credited with assembling the volume in which the Waldseemüller world map was found, was represented by his ex libris bookplate, his celestial gores, and the fragment of his terrestrial gores that had been used to strengthen the volume's binding. One of the original twelve sheets of the Waldseemüller map, usually the one demonstrating the word "America" on the southern continent in the New World, was displayed at a forty-five-degree angle in a specifically designed, hermetically sealed transparent case with a constant interior temperature of 18 to 20 degrees Celsius and a relative humidity of 50 +/–5 degrees, and with lighting maintained at a low voltage.

The fourth section, an epilogue focused on the controversies concerning the document, is extensively presented in chapters 2, 4, 6, and 7.

The door to the events leading to the consummation of the purchase by the Library of Congress was more widely opened on the occasion of the United States of America's celebration of the quincentenary of Columbus's discovery. In 1992, the map was on display in its entirety, with the twelve sheets appropriately placed side to side, as the centerpiece of the National Gallery of Art's exhibition "Circa 1492: Art in the Age of Exploration." This was the first time that the original map was hung as a unit, emphasizing its enormity and elegance.

In November 1991, on the eve of the exhibit, Count Johannes Waldburg-Wolfegg (fig. 17)—who would be designated prince on the death of his father in the fall of 1997—and his friend Ambassador William H. Fitzgerald met with Dr. James H. Billington, the Librarian of Congress, Ralph E. Ehrenberg, Chief of the Geography and Map Division, and Winston Tabb, Associate Librarian for Library Services. At that time, the count indicated an interest in selling the map specifically to the Library of Congress.[8]

In March 1992, Daniel Boorstin, the former Librarian of Congress, wrote to the count: "It gives me great satisfaction to learn from Winston Tabb that there is possibility that the Waldseemüller Map may be available for sale to the Library of Congress. This is the earliest multi-sheet printed wall map of the world to survive."[9] Boorstin had indicated that he regarded the map, the Declaration of Independence, and the Constitution as the three most important documents in the history of the United States.

Appropriately, it was just before Independence Day in 1992 when Ralph E. Ehrenberg and Margrit B. Krewson, the German/Dutch area specialist at the Library of Congress, met with the count at his

Figure 17. Prince Waldburg-Wolfegg-Waldsee and Princess Viviana at the occasion of the display of the Waldseemüller world map of 1507 during the opening of the "Rivers, Edens, Empires: Lewis & Clark and the Revealing of America" exhibition, July 23, 2003, Library of Congress, Washington, DC.

Wolfegg Castle. Prince Johannes Waldburg-Wolfegg stands out as an individual meriting appreciation for his willingness to part with a treasure that had been in possession of his family for centuries.

In the midst of ongoing negotiations with the Library of Congress, Johannes Count (later to became Prince) Waldburg-Wolfegg was interviewed in 1997 by a German periodical in reference to his role as the chief executive officer of his estate since 1993. The article pointed out that his position required the combination of "a nobleman's awareness of tradition with modern entrepreneurship. He has restructured his estate into several limited liability companies including a forest and wood industry, vineyards, a health resort, and a golf resort in addition to overseeing the family's extensive and significant art collection." The current prince adhered to the dictates that the "Corpus be kept locked in the possession of the family," as stated in the will of Max Willibald, his distinguished ancestor, who defended Constanz, Lindau, and Wolfegg during the Thirty Years' War and is credited with having purchased the Waldseemüller maps circa 1640. Therefore, the binding and its contents, the two Waldseemüller maps, and the Schöner gores were rarely exhibited publicly and, as noted earlier, they were stored in a large freestanding safe that was secured with six key locks.

Mrs. Krewson merits specific recognition as a catalyst, liaison, expediter, and persistent protagonist in the decade of discussions and deliberations that finalized the contract of transfer. She is a prime example of a government worker laboring in the background of scholarship, whose efforts define the broad dimensions and depths of the Library of Congress. Mrs. Krewson served the Library of Congress from 1967 until her retirement in 1998. The position she held had been created in 1942 for Thomas Mann, the acclaimed German author, after he had been driven out of Germany and immigrated to the United States.

For the three-year-interval between Mrs. Krewson's retirement and the library's acquisition of the map, the project remained a cause célèbre for Mrs. Krewson. Even after her retirement from the Library of Congress, she personally expended thousands of dollars to bring the acquisition to its conclusion. As the prince wrote in a letter to an administrative officer of the Library of Congress (Mr. Tabb) in July 2001, "*Die Welt*, the second-largest national newspaper in Germany, today published an article pointing out that the Library of Congress is in fact the ideal owner of the Waldseemüller Map. Its author mentions the unparalleled persistence and continuous efforts of Margrit B. Krewson to have made this possible. And I could not agree more."[10]

In the Geography and Map Division of the Library of Congress, there is a large corpus of files of correspondence, reports, and memos detailing the chronology of the processes of negotiation, the clearance of German restriction of transfer of national treasures, the efforts of raising the money required for purchase, the multiple modifications of contracts, the finalization of the sale, and the transfer of the documents to their permanent home in Washington, DC. (Much of the following is based on this large collection of letters, correspondence, reports, etc., contained in the Library of Congress archives dealing with the acquisition and transfer of the map.)

Ralph E. Ehrenberg and Margrit Krewson's 1992 visit to Wolfegg Castle succeeded in eliciting from the count an affirmation that the map would be offered solely to the Library of Congress. A price of $10 million was established, based on the evaluations of Sotheby's, Christie's, and a Geneva firm.

Looming as a significant obstacle to the transfer of the map to the United States was the intransigent posture of the German government concerning the sale. The German government under Chancellor Helmut Kohl, a historian by training, disallowed the export of what was classified as a national treasure. No item registered on the German comprehensive list of valuable national cultural property had ever been released for an export license. In 1994, the Librarian of Congress, Dr. James Billington, met with Professor Werner Weidenfeld, coordinator for German-American cooperation, who promised to "continue to press the [legal] issue," but it was apparently never brought to Chancellor Kohl's attention. At the same time, Margrit Krewson persistently explored various options with German officials at the highest levels of government and industry.

When the president of the Federal Republic of Germany Richard von Weizsäcker (1984–1994) visited the United States, Mrs. Krewson enlisted the support of the then deputy chief of Mission at the German embassy, Fritjof von Nordenskjoeld, to present the Library of Congress's request to the German president. At the behest of the president, the minister of interior, who represented the major determinant, convened a panel of experts consisting mainly of German library and museum directors. They voted overwhelmingly against the transfer.

When Roman Herzog took over the office of president, Mrs. Krewson's German friend, whose wife had written her dissertation on the Waldseemüller map, approached President Herzog on the matter. This resulted in President Herzog's request for a feasibility study pre-emptive to the possible removal of the map from the national registry. Concurrently, in the United States, Dr. Ronald Grim, specialist in cartographic history at the Library of Congress, provided presentations focusing on the map to more than 350 visiting German dignitaries in the hopes of enlisting their support.

In August 1997, anticipating the potential of the sale, Count Waldburg-Wolfegg, accompanied by his legal counsel, Dr. Alexander Goepfert of Düsseldorf, and his curator, Dr. Bernd Mayer, undertook a tour of the treasures of the Library of Congress. At the time that the Library of Congress had received the right of first refusal, the folio containing the Waldseemüller world map of 1507 was listed on the National Register of Protected German Cultural Property. Chancellor Kohl's loss to Gerhard Schröder in the 1998 national elections in Germany was a critical event in the ultimate transfer of the map to the United States of America.

Chancellor Schröder's appointment of Dr. Michael Naumann to the influential position of federal minister of culture was a fortunate choice, because Dr. Naumann, as longtime head of the Bertelsman Group (USA), was keenly aware of the importance of German-American relations. The process of obtaining permission for transfer or sale of a German treasure to another country entails a specific sequence. Each German state has its own independent cultural ministry. Thus, release of the map initially had to be approved by the government of the state of Baden-Württemberg before the federal government could issue a stamp of approval. Approval is first granted by the state's minister of culture, after which the minister-president of the state issues approval, which is then perfunctorily agreed to by the chancellor or president of the federal German government.

A letter from Mrs. Krewson to Dr. Billington, written on December 24, 1997, indicated that she had recently spoken with the count, who informed her of his meeting with the minister-president (governor) of Baden-Württemberg and his staff. Hitting a snag, the count met with the minister-president. This resulted in the agreement that if the count would donate two paintings from his private collection to the Baden-Württemberg Gallery of Art, the export permit would be granted. The count agreed to the terms, realizing that once the state issued approval, the federal government would rubber-stamp the license for export.

The formal offer was transmitted in a letter written by Dr. Jörn Günther, antiquariat, to Mr. Winston Tabb, associate librarian for library services of the Library of Congress, on May 25, 1999:

> I am representative for Prince Waldburg-Wolfegg in the negotiations with the Library of Congress for the Waldseemüller Map of America, Carta marina and globe gore fragments. The Prince has authorized me to offer the Waldseemüller Map of America, Carta Marina and globe gores to the Library of Congress for $14,000,000.
>
> The Prince understands that the Library will offer numerous duplicate books in exchange, which will be credited, after mutual evaluation, towards this $14,000,000, amount. The cash balance of $14,000,000,—, less the exchange items would be payable in full when the export of items is approved, June 14, 1999.
>
> The Prince must have an executed purchase agreement from the Library of Congress for the map materials, in hand, by May 31, 1999 in order to complete his application of these materials at Bonn June 14, 1999. Without a purchase agreement from the Library of Congress the export cannot be addressed at the June 14 meeting

and the window of opportunity that exists to place these items at the Library of Congress will be lost forever.[11]

The prince's designation of the price was based, in part, on a consultation with Philip D. Burden, an English map dealer and author of *The Mapping of North America*, published in 1996. According to Burden, the events unfolded like a spy novel. His involvement began on July 24, 1998, when he was contacted by a firm of London bookdealers and asked for help in authenticating a map for one of its clients. When told that his services were needed the following week, Burden informed the firm that, on the weekend, he was beginning a three-week vacation with his family in San Diego. He suggested that his father, Clive A. Burden, a notable map dealer who had established their firm, could substitute. But that was deemed unacceptable. When he asked what the item of interest was, Burden was informed that he would first have to sign a confidentiality agreement. Once signed, he was told, "We have a client who is interested in purchasing the only known examples of the world map of Martin Waldseemuller, 1507 and the 'Carta Marina' of 1516."

After flying to California with his family and checking in at a Disneyland hotel, Burden returned to London the next day and met with the bookdealers in the Lufthansa lounge. The group proceeded immediately to Hamburg, where they met the client at the terminal. A taxi took them downtown to a bank where the book was brought into a conference room for inspection. Burden confirmed the authenticity of the item and detailed his opinion in a written report. He returned to California after a night's sleep. Burden has offered the opinion that, after the Declaration of Independence and the Constitution, the 1507 Waldseemüller world map is the most important piece of Americana in existence and is the "birth document of America."[12]

In anticipation of the need to justify the price in order to raise money for the purchase, Dr. Ronald Grim of the Geography and Map Division prepared a memo for Winston Tabb, associate librarian for library services. The historical comparisons included: (1) In 1989, the

Hereford Mappa Mundi, a medieval manuscript drawn in 1290, was put up for auction with a valuation by Sotheby's of $11 million. It was withdrawn from sale. (2) A 1531 manuscript world map by Maggiolo was, at the time, being offered by a Paris dealer for $6.5 million. (3) The set of Waldseemüller globe gores purchased by the dealer Hans P. Kraus for $35,000 in 1960 was sold to the Bavarian State Library in 1991 and was valued at $3 million. (4) In 1990, a copy of the 1482 Ulm edition of Ptolemy's *Geographia* on vellum sold for $2 million. (5) English map dealer Philip Burden indicated that he had a customer willing to purchase the Waldseemüller world map of 1507 for $10 million.[13]

In December 1997, the Library of Congress compiled a list of books within its collection that could be used in lieu of actual dollars to obtain the map. The list included works by Alberti, Boccaccio, Herodotus, Petrarch, Euclid, as well as early bibles. Although the possibility of substituting books for dollars persisted for almost two years during the negotiations, the idea never reached fruition.

On June 14, 1999, Prince Waldburg-Wolfegg-Waldsee and Margrit Krewson met with Dr. Nevermann, deputy to Dr. Michael Naumann, federal minister of culture in the federal chancellor's office in Bonn. Emphasizing the fact that the German federal government had never previously issued an export for one of its national treasures, Dr. Nevermann urged the prince to reconsider his decision and attempt to sell the map within Germany. But, because Dr. Nevermann had received specific instructions from Dr. Naumann to grant the license, the issuance was formally consummated. The culmination of these efforts was chronicled in a letter from the prince to Dr. Billington, dated June 17, 1999, indicating

> It is with great satisfaction and joy that Mrs. Krewson and I have learned last Monday in Bonn of the decision of the German Government to grant permission for the export of the Waldseemüller Map to be acquired by the Library of Congress. . . . At this time I want to express my deepest gratitude to you personally and particularly to Mrs. Krewson for your continuous support in this matter during the past seven years. We remember well the number of obsta-

cles to our path to accomplish this goal. I believe this is due to a large extent to the enormous determination and persistence of Mrs. Krewson. Please convey my sincerest thanks to her.[14]

Shortly thereafter, the first contract of sale was drawn up by Dr. Alexander Goepfert of Düsseldorf and was presented to the Library of Congress. Several iterations were required to satisfy both parties. A particularly contentious issue centered around whether the map would be displayed in the Library of Congress in its original form "permanently" or "regularly in its original form or facsimile." Another item to be resolved concerned the responsibility for the costs of transportation and insurance for the map during transport.

The final contract stipulated a sale between "His *Serene* Highness Johannes Fürst zu Waldburg-Wolfegg und Waldsee" and "The Library of Congress" for the world map by Martin Waldseemüller, titled "*Universalis cosmographia secundum Ptholomaei traditionem et Americi Vespucii alioruque Lustrationes*," "listed under number 01301 in the comprehensive list of valuable national cultural property in Germany and . . . therefore subject to export restrictions provided by the Protection of German Cultural Property Migration Act of August 5, 1955" for the purchase price of US $10 million.

The contract specifies that the sale is conducted:

In consideration of the fact that:

- The Map contains the first known use of the name "America" as an original invention by Martin Waldseemüller to designate the new continent discovered by Christopher Columbus in the year 1492;
- The Map is the only existing copy of a woodcut made by Martin Waldseemüller, probably in the year 1507;
- The invention of the name "America" by Martin Waldseemüller for a new continent that had previously been designated as "terra incognita" bestows an historical identity upon the continent, and

- On this basis, the Map by Martin Waldseemüller represents a document of the highest importance to the history of the American people.

In consideration of these facts, the Library is interested in acquiring the Map, now in the ownership of Fürst zu Waldburg-Wolfegg, in order to make it available to researchers and the American and the general world public, and to permanently display it in the Jefferson Building of the Library, in original or—if necessitated by the Library's preservation requirements, building repair, maintenance, renovation or other circumstance requiring temporary removal of the original—for some time in facsimile.

In consideration of the fact that:

- The map is of unique importance to the history of America and the American people;
- It is in the public interest to make the Map available to researchers and the American and general public;
- The objective behind selling the Map to the Library is to enhance the cordial relationship between Germany and the United States

the Fürst zu Waldburg-Wolfegg is willing to sell the Map to the Library.

Dr. Billington signed the contract on September 14, and the prince signed it on October 13, 1999.

The Library of Congress transferred $500,000 to the prince as a down payment, and efforts were initiated to fulfill the financial commitment. Dozens of letters were generated and directed to corporations and individual philanthropists in an attempt to raise money for the purchase of the map. Eventually, the need was satisfied by three sources. In its report making supplemental appropriations for the fiscal year 2001, the House Appropriations Committee (House Report 107–102, June 20, 2001) endorsed the library's effort to acquire the map and allocated $5 million toward the purchase. A precedent for such a contribution had been established by the 76th Congress when it enacted a bill, H. R. 6990 on June 26, 1939, "to acquire by purchase, a certain locket known as the 'Castillo Locket' and a gold and crystal crucifix, each containing fragments of the dust of Christopher Columbus. . . . This is authorized to be appropriated not more than $50,000 to carry out this Act."

In announcing the purchase of the map in June 2001, James H. Billington, Librarian of Congress, stated:

> This map, giving our hemisphere its name for the first time, will be the crown jewel of the Library's already unparalleled collection of maps and atlases. The purchase marks the culmination of an effort that has extended over many decades to bring this unique historical document to America where it can be on display in the nation's library for all to see. The Library of Congress is grateful to Prince Johannes Waldburg-Wolfegg and to the governments of the Federal Republic of Germany and the state of Baden-Württemberg, which made this acquisition possible.

On June 27, 2001, Dr. Bernd Mayer, curator of the Wolfegg collection, drove from the castle to Zurich, where he boarded a Swissair flight to Dulles Airport in Washington, DC. Two business-class seats had been purchased to allow the treasure, which had been placed in a wooden box, to be maintained by the side of the courier. However, its size precluded this, and the box was placed in the hanging-clothes compartment in the cabin. The value of the material was attested to be $10 million in establishing the $7,000 fee for insurance during the flight. Awaiting the afternoon arrival of the plane were James Flatness from the Geography and Map Division of the Library of Congress, Mark Roosa from the Library of Congress, Lieutenant Murphy of library security, and the legal counsel from the Library of Congress, all of whom had arrived in a van. Also present were two representatives of the prince. Dr. Mayer transferred the box containing the volume to the van, which transported him, the security lieutenant, legal counsel, and Mark Roosa to the basement receiving room of the

Library of Congress. James Flatness and the prince's representatives followed in a large Mercedes. The box was taken to the conservation department of the library, where it was uncrated and the contents were examined and authenticated by Mark Roosa and the conservationist Heather Wanser. America's Baptismal Document had arrived at its new and permanent residence!

But the financial commitment had not been fully satisfied, and a six-month extension for payment was required. The residual requirement was finally met when Gerald Lenfest and David Koch of the Madison Council of the Library of Congress made substantial gifts, and the remaining $3.5 million was realized by a contribution from the Discovery Channel, with the understanding that the library would assist the television network in developing a series of thirty geographical/cultural programs under the general title of "The Atlas of the World." On June 18, 2003, it was announced in *NEWS* from the Library of Congress that the purchase had been completed.

On July 23, 2003, the Waldseemüller world map of 1507 was displayed for the first time in the Thomas Jefferson Building of the Library of Congress as a companion to the opening of the Lewis and Clark exhibit "Rivers, Edens, Empires: Lewis & Clark and the Revealing of America." On that occasion, Prince Johannes Waldburg-Wolfegg recounted the past refusals of his father, grandfather, and great-grandfathers to sell the map, and then went on graciously to express his personal sentiments regarding his great admiration for the United States of America: "This place, the Library of Congress, is a better repository than a castle in southern Germany. As my family kept this map . . . close to its heart, I want to place it close to the heart of the Library of Congress, to care for it as my family has for over 350 years."

Thus, the grand document defining America's baptism became the property of the United States of America. The acquisition was enhanced, shortly thereafter, when the remainder of the cartographic contents of the folio that Professor Fischer uncovered at Wolfegg Castle in 1901 joined the 1507 world map as a complementary treasure of the Library of Congress. Once again, Mrs. Margrit Krewson played a critical role in securing the "Carta Marina" of 1516, the Schöner celestial and terrestrial globe gores, and the portfolio for the library. In December 2002, while arranging an exhibit, "Treasures from the Dresden Library," at the Miami-Dade Public Library in Miami, Florida, she met Mr. Jay Kislak, a member of the library's Madison Council and a renowned private collector. She informed Mr. Kislak of the maps and offered to introduce him to the prince. The prince and Mr. Kislak met in June 2003 at a music festival held at Wolfegg Castle.

An understanding was reached between the two parties and, on November 7, 2003, Mr. Kislak acquired what for centuries had been the accompaniments of the 1507 world map. Initially, the items joined Jay Kislak's collection of pre-Columbian art and rare books, manuscripts and maps, and were exhibited as the highlight of the Miami Map Fair in February 2004. The items now reside in the Library of Congress on the second floor of the Thomas Jefferson Building, adjacent to the rare book room's collection as part of an area dedicated to Jay Kislak's art, books, and documents spanning centuries from the time of indigenous cultures through the era of exploration to the American colonial period.

CHAPTER 9
RESIDENCE AND REVERENCE

Libraries are successively the cradles and the sepulchres of the human mind.

Santiago Ramón y Cajal

The entrance of the document that put "America" on the map into its permanent place of residence was far from ceremonial. There were no press photographers on hand, no fanfare or flag waving to celebrate the event. The routine delivery to the Library of Congress was certainly inconsistent with the importance of the event. On the other hand, the location of the ultimate destination for the map that named America remains eminently appropriate. The map that assigned the name, which became incorporated in the title of a great nation, arrived at the greatest repository of that nation's historical documents. The Library of Congress, with its architectural elegance and its unequaled collection of diverse items, would forever serve as a fitting shrine for the greatest jewel of American cartography. Permanent residence in the Thomas Jefferson Building of the Library of Congress places the monumental Waldseemüller world map of 1507 in an environment that enhances its role as the graphic overture to the nation's history.

The Thomas Jefferson Building is "a national cultural showpiece —with appropriately grand and beautiful interiors . . . a fitting temple for great thoughts of generations past, present, and to be." As the earliest of the three buildings that constitute the Library of Congress, it contributes to the oldest cultural institution in the United States of America and is the largest repository of recorded knowledge in the world.[1]

Great institutions frequently are associated with an intriguing history; the Library of Congress stands as an illustrious example of this generalization. Its beginning can be specifically dated to April 24, 1800, when President John Adams signed a bill "to make provision for removal and accommodation of the Government of the United States whereby the Executive Mansion and Congress in the District of Columbia would be outfitted after the move to the new capital city was made." The bill also provided "for the purchase of such books as may be necessary for the use of Congress at the said city of Washington, and for fitting up a suitable apartment for containing them, and for placing them therein the sum of five thousand dollars shall be, and hereby is, appropriated."[2]

The first collection that was assembled for the library consisted of 152 works in 740 volumes; three maps, consisting of Aaron Arrowsmith's two maps of America and William Faden's map of South America; and four atlases, including an 1800 edition of *American Atlas* compiled by Thomas Jefferys, geographer to the king of England. The works arrived from England on the packet ship, appropriately named *American*, on May 1, 1801. On January 26, 1802, President Thomas Jefferson signed a bill "that gave the president of the Senate and the Speaker of the House the power to regulate the Library." The act also stipulated that a librarian be appointed by the president of the United States, and that the material, which was housed in the library, could be borrowed only by the president, the vice president, and members of Congress.[3]

The first library catalog, *Catalogue of Books, Maps, and Charts, Belonging to the Two Houses of Congress*, was published in 1802 and included seven maps and six atlases. Ten years later, the first classified catalog, listing 3,076 volumes and 53 maps, charts, and plans, was issued. On August 25, 1814, toward the end of the War of 1812, the British destroyed the Capitol of the United States, including the Library of Congress. Books were used to kindle the fire, perhaps as retaliation for the burning of the legislative library at York (Toronto), Canada, by American troops in April 1813.[4]

Almost immediately upon learning of this tragedy, Thomas Jefferson, at the time in retirement at Monticello, offered to sell his personal library to Congress in order to "recommence" its library. Jefferson's library consisted of more than twice the number of volumes that had been destroyed in the fire. Among the historical gems, Thomas Jefferson's library included Ortelius's atlas *Theatrum Orbis Terrarum*, the first edition of Hakluyt's *Voyages*, Herrera's *History of the Indies*, Gomara's *Conquest of Mexico and New Spain*, Garcilaso de la Vega's *La Florida*, and, specifically related to the naming of America, a copy of the *Lettera* of Amerigo Vespucci. The panoramic scope included the arts, architecture, science, literature, and geography. This formed the basis for the eventual transformation of the Library of Congress from a legislative facility, devoted solely to governmental needs, to a library that would satisfy the scholarly interests of an entire nation. At the time of the Jefferson acquisition, the privilege of using the books was extended to the United States attorney general and members of the diplomatic corps.[5]

The library's holding were housed for a short period at Blodget's Hotel, a building at Seventh and E streets that had served as a theater but never a hotel. The library was next relocated to the north wing of the rebuilt Capitol in 1818. In 1824, the library moved to the center of the west front of the refurbished Capitol. At that time, the *National Journal* referred to the newly created room as "decidedly the most beautiful, and in the best taste of any in this country."[6]

Shortly thereafter, the opportunity to acquire another copy of the *Lettera*, containing the descriptions of the four voyages by Vespucci, and Fracanzio da Montalboddo's 1507 *Itinerarium Portugallensium—*

two works specifically related to the recognition of Amerigo Vespucci as the discoverer of continental land in the New World—was lost. The books were part of the collection of Obidiah Rich and were deemed "impractical and antiquated."[7]

In 1846, the Smithsonian Institution was established "for the increase and diffusion of knowledge among men." For a period of time, the Smithsonian Institution, under the leadership of its aggressive librarian Charles Coffin Jewett, who wished to establish a national library at that institution, represented a distinctly competitive threat to the Library of Congress. This institutional competition was not formally resolved until 1866, when the Smithsonian's library was removed from that location to the Capitol and incorporated into the Library of Congress.[8]

On Christmas Eve of 1851, a fire destroyed about thirty-five thousand of the library's fifty-five thousand volumes, including "nearly all [of its] extensive collection of Maps." At the time of the investigation of the fire, the library's name was formally changed from the National Library to the Library of Congress. In 1853, the new fireproof library room was opened in the west front of the Capitol, adjacent to the rotunda. It was heralded as "the largest room made of iron in the world."[9]

During the second half of the nineteenth century, Ainsworth Rand Spofford was responsible for transforming the Library of Congress from a legislative library into an institution of national significance in the Jeffersonian spirit. Under Spofford's leadership, the library published its first alphabetical author catalog in 1864 and succeeded in effecting the transfer of the Smithsonian Institution's forty thousand volumes to the Library of Congress, thereby obviating a challenge to the basic mission of the Library of Congress.[10]

In 1867, the librarian persuaded Congress to allocate $100,000 to purchase the private library of Peter Force, which formed the foundation of the Library of Congress's Americana and incunabule collection. This collection was the most important single acquisition since the 1815 purchase of Thomas Jefferson's library. As Spofford said: "In the field of early American printed books, so much sought for by collectors, and which are becoming annually more scarce and costly, this library possesses more than ten times the number to be found in the Library of Congress."[11]

The gestational period for a building devoted solely to the Library of Congress was long. In March 1873, President Grant approved an appropriation to plan the erection of an appropriate home for the library. The building was not authorized until 1886, and the final design of a structure to be located across the east plaza from the Capitol was not approved until 1889. The new 326,000-square-foot building, which almost eighty years later would be named the Thomas Jefferson Building, was completed and ready for occupancy in February 1897. Finally, the bronze doors representing Tradition, Writing, and Printing were opened to the public on November 1, 1897.

From 1899 to 1939, Herbert Putnam occupied the role of Librarian of Congress. His tenure was noteworthy because of a series of significant acquisitions. In 1915, the collection of Henry Harrisse, the notable scholar and Americanist, came to the library. In 1927, John Boyd Thacher's collection was bequeathed to the library. It includes works produced by more than five hundred fifteenth-century presses, highlighted by a copy of the 1493 Basel edition of the Columbus letter *De insulis in mari Indico nuper inuentis*, and a copy of the 1507 edition of *Cosmographiae Introductio*. But perhaps the most notable acquisition during Putnam's tenure was the three thousand-volume Vollbehr collection of fifteenth-century books, including one of three known perfect copies of the Gutenberg Bible on vellum in existence. In 1930, at the time of the Depression, Congress approved a special appropriation that allowed for the purchase of the collection at the bargain price of $1.5 million.[12]

In 1938, the Library of Congress's large Annex Building, which was intended to serve primarily as a facility for stacking ten million volumes, was completed. In 1980, the building acquired its current name, the John Adams Building, honoring the man who approved the law that established the Library of Congress.[13] The same year, the James Madison Memorial Building also opened. It honored the fourth president, who was one of three members of the ad hoc committee of the Continental Congress that in 1783 recommended the establishment of a congressional library.

The James Madison Memorial Building, which is located directly south of the Thomas Jefferson Building, is the single-largest library building in the world, and is exceeded in size only by the Pentagon and the Federal Bureau of Investigation buildings among the government properties. It was occupied immediately by the office of the librarian, the administrative department, the copyright office, the law library, the legislative reference service, and the conservation and processing departments. It houses manuscripts, music, prints, photographs, serials, and the Geography and Map Division among its several sections.[14]

The Library of Congress, the nation's oldest cultural institution and the largest library in the world, is characterized by an immensity and diversity that challenges description. According to the library's 2004 Web site, the collection consists of more that twenty-nine million books, fifty-seven million manuscripts, twelve million photographs, almost five million maps, and three million recordings. Among this plethora are two segments, namely, the German Collection and the Geography and Map Division, that merit special consideration because of their relationship to the document that put "America" on the map.

Given the facts that the word "America" derives from an Italian name with a Germanic source; that the man who was responsible for creating the world map of 1507, which provided a name eventually adopted by the United States, was born in Germany and educated at a German university; that the volume containing the Waldseemüller world map of 1507 was collated and initially maintained by a German scientist; that the document spent almost four centuries in a German castle, where it was discovered by a German priest who wrote of his discovery in the German language; that the opportunity for the Library of Congress to possess the map was granted by a German prince; and that the negotiations were catalyzed by a specialist in the

German/Dutch section of the library, the German collection of the library merits special consideration.

In 1993, Margrit Krewson—who played a significant role in the acquisition of the Waldseemüller world map of 1507 and also the remaining contents of the volume discovered at Wolfegg Castle—presented a chronology of the development of the library's German collections. At the time of that report, the German material in the library's general collections consisted of about 2,250,000 volumes, and there had been a recent annual increment of 30,000 volumes. In addition, the German language items in the Library of Congress's special collections numbered more than four million pieces.[15]

The history of the German collection began with the acquisition of Thomas Jefferson's library in 1815. In 1898, Prince Otto von Bismarck died, and the library purchased a large amount of material related to his career.[16] In 1904, the library received its first phonograph record, a cylinder recording of the voice of Kaiser Wilhelm II. The same year, the library purchased the library of Albrecht Weber, professor of Sanskrit at the University of Berlin. Consisting of 3,018 books and 1,002 pamphlets, the collection contained "the foundation for all work in Indian philology." At about the same time, the library acquired the Kohl Collection of manuscript maps.[17]

During World War I, the library purchased a pristine 1790 copy of Goethe's *Faust* and eight full-page woodcuts attributed to Albrecht Dürer. Between the wars, in addition to the Vollbehr collection, which included the Gutenberg Bible, the library acquired from Otto H. F. Vollbehr the William Schreiber Collection, which contained more than two hundred thousand book illustrations from the fifteenth- to eighteenth-century as well as publications relating to the Reformation by Martin Luther and John Calvin. In 1928, the papers of Baron Friedrich Wilhelm von Steuben, who helped train troops at Valley Forge, were purchased from the New York Historical Society, which added to the library's collection of the papers of George Washington. After World War II, the library acquired books belonging to Adolf Hitler, Hermann Göring, and Heinrich Himmler.

The extraordinary Lessing J. Rosenwald Collection, which was presented to the library between 1943 and 1964, includes more than five hundred books printed in the fifteenth century. It is believed that one of the items, the 1460 *Catholicon*, was printed by Johannes Gutenberg. The gem of that collection is undoubtedly the fifteenth-century manuscript Giant Bible of Mainz, now located in the same area of the Thomas Jefferson Building as the Gutenberg Bible.

The Library of Congress's Sigmund Freud Collection contains over eighty thousand items, including manuscripts of his books and papers. In the musical realm, the Library of Congress houses the Johannes Brahms Collection, which contains more of his manuscripts and related works than any institution outside of Vienna. The library's Whittall Foundation Collection boasts manuscript sketches and scores of Johann Sebastian Bach, Ludwig van Beethoven, Alban Berg, Joseph Haydn, Felix Mendelssohn, Giacomo Meyerbeer, Wolfgang Amadeus Mozart, Arnold Schoenberg, Franz Schubert, Richard Wagner, and Carl Maria von Weber.

Daniel J. Boorstin, the twelfth Librarian of Congress and former director of the National Museum of American History, was a Pulitzer Prize winner, a distinguished professor of history at the University of Chicago, and a popular author. When asked to compose a list of his preferences among the millions of items contained by the Library of Congress, he included the Great Hall of the Thomas Jefferson Building, the Giant Bible of Mainz, and the Geography and Map Division.

It is appropriate that the Geography and Map Division of the Library of Congress occupies center stage in the final act of the drama of the Waldseemüller world map of 1507. Although two maps of America by Aaron Arrowsmith, a map of South America by William Faden, and four atlases, including an 1800 edition of the *American Atlas* compiled by Thomas Jefferys were part of the nucleus of the library's collection in 1800, it was not until the library opened its building in 1897 that a separate section was created to oversee the map collection.[18]

During that interval, a few significant acquisitions were made by the library. The earliest was the William Faden Collection that includes 101 manuscript and printed maps related to the French and Indian War. The collection was purchased for one thousand dollars from Edward Everett Hale, who later became chaplain to the US Senate and is best remembered for his tale *The Man without a Country*. The 1867 purchase of the Peter Force Collection, which contains more than twelve hundred maps and views, significantly augmented the library's cartographic holdings. In 1882, the library acquired twenty colored manuscripts of New Jersey drawn by John Hill, an engineer on the staff of Sir Henry Clinton, British Supreme Commander in America. The next year saw the purchase of a collection of forty manuscript maps, a manuscript atlas, and twenty-eight printed maps belonging to the Comte de Rochambeau, the commander in chief of the French army in America during the Revolutionary War.[19]

At the time that the Library of Congress moved from its quarters in the Capitol to the spacious new building, Philip Lee Phillips, who had joined the library's staff in 1875, was appointed superintendent of the Hall of Maps and Charts. At its inception, the hall contained approximately twenty-seven thousand maps and twelve hundred atlases. Between 1909 and 1920, the first four volumes of the *List of Geographical Atlases in the Library of Congress* compiled by Phillips were published.[20]

The Hall of Maps underwent a series of name changes: first, Division of Maps and Charts, then Division of Maps, and then Map Division, until 1965 when it received its current name, the Geography and Map Division. In 1957, the division moved to the Annex Building. Twenty years later, the division moved to an annex in Alexandria, Virginia, where it had temporary quarters until 1980, when it finally settled in its current location in the basement of the James Madison Building.

The Geography and Map Division houses and has custody of the great majority of the library's cartographic material, but some of the treasures are intermittently or permanently displayed in the Thomas Jefferson Building. The Library of Congress, with the unrivaled elegance of the Thomas Jefferson Building coupled with the space allocation and functionality of the adjoining James Madison Building, provides the environment that includes the nation's largest conservation facility and incorporates within its walls preservation and storage, scholarly activities, educational endeavors, and captivating displays. The buildings serve as an exoskeleton for the functioning elements—the atlases, maps, and globes—an unequaled source of information and history.

The Geography and Map Division's collection encompasses a broad expanse of history. Among the printed works, the holdings begin chronologically with the works associated with Claudius Ptolemy, the second-century Alexandrian scholar. His material was discovered by Renaissance Europeans and constituted the first atlases to be printed with the newly developed movable type in the fifteenth and early sixteenth century. The library possesses forty-seven of the fifty-six editions of Ptolemy, which date from 1475 to 1883.

The library's sixteenth-century holdings include a rare collection of maps assembled by Antoine Lafrery, who perfected the copperplate engraving of maps; Abraham Ortelius's 1570 *Theatrum Orbis Terrarum* (Theater of the World), which presented for the first time several maps of uniform size and design in a book; and the 1595 *Atlas sive cosmographicæ meditationes de fabrica mundi et fabricati figura* by Gerhard Mercator, who introduced the word *atlas* to honor the Titan Atlas—the mythical king of Mauritania, philosopher, astronomer, and mathematician. Also included in this time span are atlases of England and France, and the first atlas devoted exclusively to the New World, Corneille Wytfliet's 1597 *Descriptionis Ptolemaicæ augmentum, sive Occidentis*.[21]

A wide variety of seventeenth- and eighteenth-century Dutch, French, English, German, and Italian world atlases are to be found on the shelves and in the vaults under controlled environmental conditions. The library's collection of printed sea atlases is enriched by thirty-three mid-fourteenth- to eighteenth-century portolans. They are found in the vault in the Vellum Chart Collection.[22]

Atlases related to the French and Indian War, the Revolutionary

War, the Civil War, the two world wars, and the more recent engage-
ments in which our nation participated all provide a graphic history of
those events. The United States' entry into the business of atlas publishing at the end of the eighteenth century is well represented. Maps of
regions, states, and cities bring the collection up to the most recent time.
Thematic maps, which include those presenting railroad routes, fire
insurance maps, physical geography atlases, agricultural atlases, cultural
atlases, economic atlases, and geological atlases depict the evolution of
the nation and provide an element of modernity in the collection.

When the Waldseemüller world map of 1507 became a part of the
cartographic holdings of the Library of Congress, it immediately
ascended to the status of the preeminent treasure among many extraordinary treasures. The enormity of the treasure trove that is part of
the Library of Congress precludes an all-inclusive consideration but
necessitates selective discussion of some of the cartographic gems that
the Waldseemüller map has joined.

The collection created by Dr. Johann Georg Kohl, consisting of
474 hand-drawn copies of manuscript and rare printed maps in European archives, depicts America from pre-Columbian times to the
mid-nineteenth century. The collection of Woodbury Lowery focuses
on the Spanish possessions in the United States covering the period
from 1502 to 1820. The bequest made by the noted bibliographer and
historian Henry Harrisse includes the so-called Manatus map of 1639
by Joan Vingbooms, cartographer to the prince of Nassau, for the
West Indian Company of Holland. That map "Drawn on the Spot" is
the earliest depiction of Manhattan. Also part of the gift was Samuel
de Champlain's chart of his exploration of the northeast coast of
North America from Cape Sable to Cape Cod, which he drew on
vellum and dated 1607.

Among the many gifts to the Library of Congress by the distinguished collector Lessing J. Rosenwald are more than five hundred
rare books, including more than two hundred incunabula (works
printed before 1500). He enriched the library's cartographic collection of sixteenth-century maps with five maps that are among the

rarest and most significant cartographic documents related to early American history. Diego Gutiérez's 1562 "Americae sive qvarte orbis partis nova et exactissima descriptio" is one of two known copies. It has the distinction of being the largest map of the New World of its time and the first to contain the name "California," applied as "C. California" to the lower tip of the Baja Peninsula. André Thevet's 1581 "Le Novveav Monde Descovvert et Illvstre de Nostre Temps" is also one of only two known copies. Franz Hogenberg's 1589 "Americae et proximarvm regionvm orae desciptio" and "Vniversale descrittione di tvtta la terra conoscivta fin qvi," executed by Gastaldo and Fernando Bertelli in 1565, are also extremely rare; only one other copy of the latter has been discovered. No more than four copies of Gabriel Tatton's "Maris Pacifici," one of the most elegantly engraved early maps depicting America, have been identified.[23]

But, the most elegant and significant gift from Lessing J. Rosenwald to the Library of Congress was the Giant Bible of Mainz, a sumptuous two-volume illuminated manuscript that was produced in 1452–1453. It is currently on display in the Great Hall of the Thomas Jefferson Building across from the Gutenberg Bible, offering a most appropriate segue from manuscript to print.

The Waldseemüller world map of 1507 chronologically leads a most impressive roster of early maps that depict the discovery, settlement, and expansion of the land, which would evolve into the United States of America. Drawing at random from the collection could bring forth John Smith's 1612 map of Virginia and his 1616 map that named "New England"; one of the five known copies of Augustine Hermann's 1673 map of Virginia and Maryland; John Foster's 1677 woodcut of New England, the first map printed in colonial America; the first maps of each and every one of the thirteen original English seaboard colonies; the only known copy of Bernard Romans's 1774 large map of Florida; and nineteen original copies of the twenty-one impressions of John Mitchell's "A Map of the British and French Dominions in North America." This 1755 map was referred to by Colonel Lawrence Martin, chief of the Geography and Map Division, in his biographical sketch of John Mitchell for the *Dictionary of American Biography*: "Without serious doubt Mitchell's is the most important map in American history."[24]

Although the Mitchell map was specifically designed to stake out the English claims on the North American continent on the eve of the French and Indian War, its major importance is related to the fact that a copy of the map was used by each of the American negotiators, John Adams, Benjamin Franklin, and John Jay, and also by Richard Oswald, the representative of King George III, in establishing the boundaries of the United States at the end of the Revolutionary War.

The library's collection of eighteenth-century original manuscripts and printed maps related to the French and Indian War and the American Revolution is unrivaled and is supplemented by photocopies of items from other libraries in the United States and the United Kingdom. As would be anticipated, the Geography and Map Division houses all of the pictorial representations of the nation's capital, highlighted by the recently restored manuscript of Pierre Charles L'Enfant's original plan for the capital city, including Thomas Jefferson's handwritten editorial changes.

The exploration and expansion that occurred in the nineteenth century is richly represented in the library. The Lewis and Clark Collection—items from which were displayed in 2003 on the occasion of the bicentenary of the overland exploration from the Mississippi River to the mouth of the Columbia River—served as an accompaniment to the initial unveiling of the Waldseemüller world map at the Library of Congress. Survey-based state maps that were drafted, engraved, printed, and published in the United States sequentially chronicle the incorporation of the contiguous territories and their entrance into the Union.

The papers of Jedediah Hotchkiss, a topographical engineer in the Confederate army, are highlighted by a collection of manuscript field sketches, regional maps, and battle plans that were prepared specifically for General Thomas J. "Stonewall" Jackson's campaign of 1862 in the Shenandoah Valley. Those maps are a significant but small segment of the most complete assemblage of pictorial material stemming from the Civil War.[25]

The twentieth century witnessed the greatest expansion in the number of cartographic items that have been made available for incorporation into the Geography and Map Division. Most recently, more than two hundred thousand aerial photographic and remote-sensing images derived from reflected energy collected by orbiting satellites have been added to the library's holdings.[26]

The basement level of the James Madison Building with its functional capabilities provides space for storage, reference facilities, and research endeavors relating to a broad expanse of cartographic concerns. Most of the items are in the general collections, but some of the more valuable material is stored in a five thousand-square-foot vault in an environment of controlled temperature and humidity. The prime jewel of the Library of Congress's incomparable holdings—the Waldseemüller world map of 1507—is now located in a temple of scholarship with an architectural artistry that is appropriate for the importance and significance of the treasure housed within its walls. The criteria for such a shrine are well satisfied by the Thomas Jefferson Building of the Library of Congress.

After a decade of renovation, the Thomas Jefferson Building was formally reopened to the public on May 1, 1997. Bounded by Independence Avenue, First, East Capitol, and Second streets, the building appears to be an extension of the Capitol separated by a street. The library, with its main entrance on Independence Avenue, faces due west. It is 470 feet long and 340 feet deep, and is made up of 409,000 cubic feet of granite, 500,000 enameled bricks, 22,000,000 red bricks, 3,800 tons of steel and iron, and 73,000 barrels of cement.[27]

The library consists of three stories—a basement, first story, and second story rising to a height of sixty-four feet. Added to this, the base and balustrade bring the total height to seventy-two feet above the ground. Located above the centrally placed octagonal reading room is a dome, the apex of which is 194 feet above the ground. The main entrance is approached on the main floor through one of three arches. A flight of granite stairs rises from the north and south sides to a central landing from which a broad stairway ascends to the entrance porch.

The main entrance beyond the three arches is dramatically presented by three massive bronze doors, each bearing a sculptured bas-relief. To the left, the door known as "Tradition" illustrates oral communication, the method by which knowledge was originally spread. The right door, titled "Writing," allegorically presents Truth and Research being conveyed by the handwritten word. Four figures, an Egyptian, a Jew, a Christian, and a Greek represent the four groups who were most influential during the time that preceded printing. Appropriately for a modern library, the center door is "The Art of Printing," on which Minerva, the goddess of learning and wisdom, is shown disseminating the products of the typographical art.

Currently, in a time of heightened security, the Thomas Jefferson Building is entered through the west ground level under a granite arch. The path from the entrance to the destination where the Wald-seemüller map can be viewed is truly awesome and in keeping with the drama of the experience. Ascending the stairway brings the visitor to the first floor's main entrance hall, a breathtaking site of beauty composed of a patterned marble floor, white marble and stucco columns replete with adorning figures, brilliant color and gilt coves, and richly embellished domes.

The west corridor on the library floor serves as a vestibule and is the most ornate, with a paneled ceiling embellished with gold ornamentation. The grandeur of the architecture is perhaps best appreciated when standing in the center of the Great Hall. The ceiling, seventy-five feet above the floor, contains stained glass skylights and elaborately paneled beams. On the east side of the hall, the commemorative arch bears the words "LIBRARY OF CONGRESS." The east corridor of the main floor is highlighted by two of the library's great treasures, one of the three known perfect copies of the Gutenberg Bible on vellum and the Giant Bible of Mainz.

Before proceeding to the second floor and the chamber that will enshrine America's Baptismal Document, it is appropriate to visit the core and raison d'être of the library, the Rotunda Reading Room. The Rotunda Reading Room was formerly entered through the center of the east side of the east corridor and is now entered from the opposite side of the octagon. The room is best viewed from the second-floor public gallery, and words cannot do justice to its grandeur and its artistic intricacies. The combination of colors, stained glass, and sculptures adorning a striking architecture create what is arguably the most attractive and elegant room in the nation. This room is the heart of the library and an unmatched sanctuary for scholarly pursuits based on researching the past.

The second floor of the library can be reached by ascending a grand stairway on either side. On the side of the right stairway is a buttress containing two putti, or cherubs—"America" and "Africa"—sitting beside a globe showing those continents. The American cherub is bedecked with a tall feathered headdress and a wampum necklace. Europe and Asia are represented by equivalent figures on the other stairway.

With the specific goal in mind of viewing of the Waldseemüller world map of 1507, the visitor should proceed to the southwest pavilion, which has been selected as the site for permanent display. Initially, a location in the area of the Gutenberg Bible was considered most appropriate because of the mapmaker's German heritage and the fact that the map is the oldest printed wall map. But that location presents problems related to the natural light entering the area, and, more important, the location is used for receptions and therefore subject to more potentially injurious traffic.

The southwest pavilion, known as the Pavilion of the Discoverers, is entered through the southwest gallery that contains a changing exhibit titled "America's Treasures." As the name implies, the exhibit displays representatives of the most important items in the collection of the Library of Congress, thereby providing a fitting "red carpet" entrance to the room that enshrines the map.

When one views the room that houses the map, one gets the sense that it is almost as if the room had been planned in anticipation of receiving the map. The lunettes (small half-moon–shaped area above the main wall) of the Pavilion of the Discoverers, beginning on the east side and proceeding to the right around the room, depict Adventure, Discovery, Conquest, and Civilization in an obviously logical sequence. The ceiling disc offers an artistic representation of the four characteristics associated with each of these stages, namely: Courage, Valor, Fortitude, and Achievement.

In the lunette on the east side, Adventure is shown as a female figure sitting on a throne, bearing a sword in her right hand while her left hand rests on a caduceus, the emblem of Mercury—the god of travelers and adventurers. At either side of the throne is a shield on which a Viking ship with oars and sail appears. To the left of the central figure, a female figure holding a cutlass in a her right hand and a pile of coins with her left hand represents the English pirates, who contributed to early adventures in the American regions. A list of English explorers who contributed to the early history of America is presented as an accompaniment. On the right side, the names of Drake, Cavendish, Raleigh, Smith, Frobisher, and Gilbert appear. That list is balanced to the left of the central figure by a list of Spanish adventurers who played roles in American adventures: Dias, Narváez, Coello, Cabeza, Verrazzano, Bastidas. The list is accompanied by the painting of a female figure holding a battle-ax and small gold figures, emblematic of Spanish interest.

Discovery, the theme of the lunette on the south side of the pavilion, is represented by a female with a ship's rudder in her right hand while her left hand rests on a globe. The globe presents the Western Hemisphere with America and was copied from a world map that had been ascribed to Leonardo da Vinci. The original map was found in a collection of papers written by Leonardo, and is preserved in the Royal Collections at Windsor Castle. R. H. Major, who first called attention to the map, credited it to the great artist. But Marquis Girolamo d'Adda has disproved that contention. Harrisse argues that the geographic representation and nomenclature indicate that the map was produced after 1519, the year that Leonardo died, and that, although the map was found among his papers, it never belonged to him. The map is regarded to be based on an early Spanish map modified by a cartographer who was acquainted with the *Cosmographiae Introductio* and/or the Waldseemüller world map of 1507 because the word "America" appears.[28]

The shield at the foot of the throne includes an astrolabe—an old device used by navigators to determine the altitude of the sun in order to compute latitude. The central figure on the lunette is flanked by two figures. The figure to the left holds a paddle and a chart and points to the distance, while the figure to the right holds a sword and a cross staff, another device to measure the altitude of the sun. On the left side, the names of Solis, Orellano, Van Horn, Oiedo, Columbus, and Pinzon appear, balanced on the right side by Cabot, Magellan, Hudson, Behring, Vespucius, and Balboa. The Discovery panel serves as the background for the Waldseemüller world map.

The lunette on the west side, Conquest, depicts a central group after a triumph bearing the insignia of victory. On one side, the figure bears a sheaf of palms, representing southern conquests, while on the right, oak leaves wreathing the figure's sword are symbolic of northern conquests. The names appearing on the left side are Pizarro, Alvarado, Almagro, Hutten, Frontenac, and de Soto, matched on the right side by Cortés, Standish, Winslow, Phipps, Velásquez, and de León.

The fourth side is devoted to Civilization. No armor is shown, and the three figures are adorned in classical costumes. The central figure is crowned with a laurel wreath and carries the torch of learning and an opened book. The figure to the left symbolizes Agriculture, holding a scythe and a sheaf of wheat. To the right, the figure representing Manufacture has a spindle for which she is twisting the thread. The names to the left are Eliot, Calvert, Marquette, Joliet, Ogelthorpe, and Las Casas, balanced to the right by Penn, Winthrop, Motolinia, Yeardley, and La Salle.

The Waldseemüller world map is displayed in its thirty-four-square-foot magnificence in a fitting case provided by a gift from Virginia Gray, honoring her late husband, Martin Gray. The hermetically sealed case has been constructed according to the standards used in the National Archives for the cases that house the manuscripts of the three most important documents in the history of the United States of America—the Declaration of Independence, the Constitution, and the Bill of Rights. The great map is still waiting to be permanently installed, and it is planned that the encased map will be flanked by two magnificent globes. Each globe measures 110 centimeters (43 inches) in diameter and stands nearly 2 meters (6 feet) high. The terrestrial and celestial globes were produced by Marco Vincenzo Coronelli in 1688 and 1693, respectively, and constitute the only complete pair of Coronelli globes in North America.

The Pavilion of the Discoverers creates a strikingly appropriate milieu for the prime cartographic treasure of the Library of Congress. The map that named America has found its permanent place of rest and reverence surrounded by the names that evoke memories of significant discoveries in the New World, including, most specifically, Columbus and Vespucci, respectively, the discoverer and the definer of that New World.

ADDENDUM

It is now likely that the Waldseemüller world map of 1507 will have a venue in the Thomas Jefferson Building of the Library on Congress that differs from that originally proposed and described in detail in chapter 9 of this work. Currently, it is proposed that the map will be displayed in the northwest pavilion, known as the Pavilion of Art and Science. The Waldseemüller world map of 1507 will share space with the Jay Kislak Collection that focuses on the European arrival to America and includes: the Carta Marina of 1516 by Martin Waldseemüller, the fragments of Johann Schöner's terrestrial globe gores, Johann Schöner's celestial globe gores of 1517, a 1482 Ptolemy atlas, a 1508 Johannes Ruysch atlas, a 1512 Johannes Stobnicza atlas, a 1520 Peter Apian[us] world map, and a copy of *Cosmographia Introductio*.

ESSENCE AND EPILOGUE

As Geography without History seemeth as carkasse
without motion, so History without Geography wandereth
as vagrant without a certaine habitation.

John Smith

The most recent decades have been characterized by an information explosion and technological advancements that are without precedence. Consequently, the modern mind has had to absorb developments occurring with such speed that undivided attention is required. Synaptic pathways within the brain rapidly progress in a forward direction in order to incorporate new information, which at some times consists of exacting refinements and at other times, truly large leaps. The future becomes incorporated into the present with such rapidity that it often precludes our consideration of the past.

The same feeling of awe that is generated by current and anticipated accomplishments should be evoked by the events that brought us to the current state. It is not unreasonable to suggest that many past discoveries merit even greater recognition when we consider the relatively primitive and uninformed environment in which they developed. The United States of America—in order to have reached its preeminence in world affairs—had to be first discovered and then settled. The focus of this narrative is the graphic chronicling of that discovery and early settlement, and the effect that one large and elegant pictorial document, a single map, had in offering to the Renaissance world a revolutionary concept of geography and a name that would be adopted by those united states.

Current technology allows anyone to stand at a given spot on the planet with a small handheld device and determine precisely within seconds the three-dimensional coordinates of latitude, longitude, and elevation above sea level. Technology has also created a society that increasingly relies on graphic information. Television crews and photojournalists provide images from around the world that inform and expand our knowledge. Graphics, generated by photo imaging and transmitted by cable or satellite, are available for many applications, including the field of cartography.

Photographs taken by orbiting astronauts and satellites offer vivid graphic representations of the land that are astounding. But are not the past pictures based on visual reports of sixteenth-century explorers and drawn by hand, albeit less precise, equally, if not more, amazing? The graphic representations, which were generated half a millennium ago, at a time when longitude could not be determined at sea and the tools of measurement were minimally refined, provided the early maps of the newly discovered land. These charts were replete with errors and misconceptions, but they also, surprisingly, depicted a geography that remains recognizable to modern inhabitants of the land.

The early maps satisfied many needs. They reported findings, excited interest, guided future exploration, established claims, verified truths, and destroyed myths. The earliest maps of a region can be credited as having seminal effects. Among the earliest maps, one map stands out among all maps in reference to the United States. The Waldseemüller world map of 1507 certainly merits reverential attention based on its priority, uniqueness, and impact. The world map of

1507, as a document, also deserves recognition for its position of eminence in the field of sixteenth-century cartography. And the importance of the map is further amplified and romanticized by its own intriguing history and associations. These associations include a panoply of diverse items and notable individuals who played major roles or made cameo appearances in the drama of the document that put "America" on the map.

It all began at an improbable site, with a mapmaker whose reputation had not been established, and with his fortuitous acquaintance with a recently written series of letters that reported erroneous material. At the base of the Vosges Mountains in the Duchy of Lorraine, in the town of Saint Dié, the mid-fifteenth-century invention of movable type that greatly facilitated the process of printing provided an impetus for the reigning Duke René II of Lorraine, at the onset of the sixteenth century, to assemble a small group of scholarly men, known as the Gymnasium Vosagense, with ecclesiastic and secular backgrounds to collate and disseminate scientific and geographic knowledge. Among the group, one priest with a particular expertise in geography, cartography, and printing, by the name of Martin Waldseemüller, played a dominant role in the process by which America received its name.

That naming process consisted of the publication of a narrative, *Cosmographiae Introductio*, and two accompanying maps on which a Christian name was imprinted on land recently discovered in the Western Hemisphere. The date that that "America" first appeared in print was April 25, 1507.

As the sixteenth century unfolded and the scholars convened at Saint Dié, Renaissance Europeans were savoring the discovery of the New World. No more than fifteen years had passed since Christopher Columbus discovered land by sailing west across the Atlantic Ocean. Columbus's four voyages extended from 1492 to 1504. In addition to landing on and sighting many of the Caribbean islands, he set foot on the northern shore of South America, on the Paria Peninsula, and, also, at several points along the east coast of Central America. But throughout his life, Columbus remained convinced that the islands he reached and the land that he had set foot on were part of Asia rather than a new continent. Six years after Columbus's first voyage, John Cabot discovered land in the northern half of the Western Hemisphere. Cabot, like Columbus, believed that he had discovered an extension of Asia.

In addition to the discoveries of Columbus and Cabot, several other explorations in the Western Hemisphere served as the basis for the geography as represented on the Waldseemüller world map of 1507. Most pertinent was the voyage conducted in 1499 under the leadership of Alonso de Hojeda. Amerigo Vespucci sailed on that expedition, which included a landing on the north shore of Venezuela a year *after* Columbus landed in the same region and which was misrepresented in print as occurring in 1497. Early in the sixteenth century, several additional expeditions to that coast, which was referred to as the "Pearl Coast," were made. During the same fifteen-year span between the initial discovery of land in the Western Hemisphere and the making of the Waldseemüller world map of 1507, Gaspar Corte-Real sailed to Newfoundland, Vicente Yáñez Pinzón discovered northern Brazil and the mouth of the Amazon River, and Pedro Álvares Cabral coasted along the east coast of Brazil. Not one of these explorers concluded that a new continent, located between Europe and Asia, had been discovered.

The earliest extant map of the New World is the manuscript by Juan de la Cosa that bears the date of 1500, but the inclusion of several geographic elements suggests a later completion of the Western Hemisphere portion, probably about 1510. The map had no influence on the thought processes of the scholars at Saint Dié. By contrast, the "Cantino Planisphere" of 1502 and a Portuguese manuscript map drawn by Nicolo Caveri in 1504–1505 are both thought to present geographic depictions that greatly influenced the expression made on the Waldseemüller world map of 1507.

During the first decade of the sixteenth century, two works appeared in print, one consisting merely of four sheets and the other of fourteen to sixteen sheets, depending on the edition, which named a previously unknown mariner as their author. There is no question that the scholars in Saint Dié had access to these narratives and that these two narratives conjointly constituted the prime basis for the naming of America. In the first of these works, *Mundus Novus* (appendix I), published in 1503 as a letter from "Albericus Vesputius," the author describes a rich and idyllic land inhabited by intriguing natives with enticing sexual activities that undoubtedly caught the readers' interest. But the major thrust of the book resided in its title, which, for the first time, introduced the words "New World" in print.

The second book appeared about two years later, originally in Italian, and was titled *Letttera di Amerigo Vespucci delle isole nuovamente trovate in quattro suoi viaggi* (Letter of Amerigo Vespucci concerning the isles newly discovered on his four voyages). The most critical so-called first voyage took place, according to the *Lettera*, from May 10, 1497, to October 5, 1498, and included a landing on the Paria Peninsula of Venezuela a year *before* Columbus. This printed error assigned to Vespucci priority for discovering continental land in the New World. The text also included the monumental declaration that the new land consisted of a large continental mass *surrounded by water and interposed between two large seas separating western Europe and eastern Asia.* This letter of Amerigo Vespucci found its way to the halls of scholarship at Saint Dié, where it was translated into Latin and incorporated into the *Cosmographiae Introductio,* the first work to come off the press of the local Gymnasium Vosagense on April 25, 1507. The misdated first voyage included in the Vespucci *Lettera* was the stimulus for the scholarly group's geographic concept of the New World. Vespucci, the purported author of the two reports, was thus elevated to heroic status, and his name was assigned to the newly discovered continental land. The name "America" appeared in the text of *Cosmographiae Introductio,* where the reason for its selection was presented, and the word also appeared in the margin of the same page, as if to emphasize its importance. Perhaps of more significance, the name "America" was placed on the southern continent in the New World on a set of gores to be affixed to a globe and also on an elegant printed wall map that was unprecedented in size.

Martin Waldseemüller, the main author of the narrative and the man who created the maps, apparently later appreciated that he had erred in honoring Amerigo Vespucci. He removed the variant of Vespucci's Christian name on all of the maps that he subsequently produced. The circumstance is analogous to modern journalism in that an emblazoned headline is inadequately refuted by a retraction that is less apparent. Other cartographers persisted in using the term, and its permanence was likely sealed when it was assigned to both northern and southern continents in the New World by the distinguished Gerhard Mercator.

The acceptance of the printed evidence that Amerigo Vespucci was the first to land on a continent in the Western Hemisphere, and that he was responsible for the introduction of the concept of a New World surrounded by the seas, was the basis for the scholars at Saint Dié to name a newly discovered continent for him. More than two centuries after the name "America" was imposed, manuscript letters by the honoree came to light. The author of the letters made no claim to priority of discovery of continental land and presented the correct date of his voyage as taking place a year *after* Columbus had landed. It is possible that the misrepresentations in the printed documents were the product of errors or deliberate embellishments by another writer or the publisher to gain wider readership. During their lives, Christopher Columbus and Amerigo Vespucci regarded each other cordially, and there is no evidence of any animosity on the part of either man toward the other.

But before the manuscript letters came to life, vitriol spewed forth from the pens of those who took the printed matter as evidence of an attempt on the part of Vespucci to usurp Columbus's glory by making a false claim of discovery. Toward the end of the sixteenth century, Bishop Bartolomé de Las Casas set in motion an attack on Vespucci that was perpetuated in the seventeenth century by the historiographer of the king of Spain. The criticism of Vespucci as a self-aggrandizer was continued into the nineteenth century by Ralph Waldo Emerson and eventually received the humorous attention of

Ogden Nash, who pointed out that it was more effective to be a promoter than the actual discoverer.

The narrative printed at Saint Dié was well received and led to an anticipation of the discovery of the two maps called for on the title page and in the text of *Cosmographiae Introductio*. In about 1890, almost four hundred years after that book appeared, the first copy of the gore map with the name "America" imprinted on South America was discovered. That map eventually has come to reside in the James Ford Bell Collection at the University of Minnesota Library. Three other copies are known to exist, one in Munich, another in Offenberg, and the third recently sold at auction.

At the beginning of the twentieth century, the plane, or flat map, called for on the title page and in the narrative of the 1507 *Cosmographiae Introductio*, remained undiscovered. It was the holy grail of American cartography, and the quest for the map was intensified by many clues that spoke of its existence. It was mentioned in letters of the map's cartographer and by a notable historian, Trithemius, who reported that he had actually purchased a copy. It was also specifically referred to on a 1510 manuscript map by Henricus Glareanus.

In 1901, Joseph Fischer, a scholarly priest who taught history and geography at the Stella Matutina College in Feldkirch, Austria, was focused at the time on research regarding the Norse voyages to North America in the tenth and eleventh centuries and began investigating the collection at Wolfegg Castle in Baden-Württemberg. The castle dates back to the twelfth century and has served as the residence or seat of the counts and princes of Waldburg-Wolfegg-Waldsee.

In the archives of Wolfegg Castle, Professor Fischer came upon the volume that contained the Waldseemüller world map of 1507. It was indeed most fortunate that the discovery of the volume, with its cartographic contents, was made by a man who recognized the nature and importance of the documents. Father Fischer immediately appreciated that he was viewing the lost Waldseemüller map and reported its discovery in an extensive publication.

The leather binding encompassed a literal treasure trove that included two works by Martin Waldseemüller, the twelve sheets of the world map of 1507, and another twelve sheets, which constitute Waldseemüller's 1516 "Carta Marina." The volume also contained the ex libris of Johann Schöner, a notable scientist and cartographer from Nuremberg, gores for a 1517 celestial globe by Schöner, and a few tattered segments of gores for a 1515 globe by the same mapmaker. The other item found within the same binding was the southern portion of a star chart, which had been engraved by Albrecht Dürer in 1515. Fischer postulated that the work of Dürer probably was the main stimulus for the purchase of the volume by a count of Wolfegg Castle.

The Dürer map chart was removed from the binding and placed in another part of the Wolfegg Castle's collection. The remaining material that shared space in the volume with the world map of 1507 is also of great importance. The fragments of the Schöner terrestrial gores of 1515 include the word "America," making it the second-earliest printed graphic document to present the word. The only extant copy of the 1516 "Carta Marina" is the earliest printed sea chart and is second only to the 1507 map as the earliest printed wall map. It includes Waldseemüller's name as its author; it dates the map; and, most intriguingly, it indicates that one thousand copies of the 1507 world map were printed. The "Carta Marina" offers an abundance of geographic information and assigns a myriad of place names that attests to the expansion of knowledge that ensued during the nine years between the production of the two great maps. The adornment of both Waldseemüller maps, more particularly, the "Carta Marina," which is replete with pictures of kings, animals, chimera, monsters, and scenes of diverse activities including cannibalism, place the documents in the realm of artistry.

The history of the Waldseemüller world map of 1507 from the time of its production to the time of discovery cannot be defined with certainty. Although it is known that its companion narrative, *Cosmographiae Introductio*, came off a recently established press at Saint Dié as two editions, each with two printings, it is generally believed that the dimensions and elaborate nature of the 1507 and 1516 maps required a larger and more sophisticated press and more experienced personnel. It is probable that the 1507 large world map and gores, and the 1516 "Carta Marina," were printed in nearby Strasbourg by Johann Grüninger.

After the sheets constituting the Waldseemüller map and its companions left the press, the presence of the ex libris of Johann Schöner on the inner portion of the binding and the inclusion of his celestial and terrestrial gores and the 1516 "Carta Marina" suggest that Schöner owned the items, and that sometime after 1516 he assembled the volume. The date that the volume was acquired by the occupants of Wolfegg Castle and from whom it was acquired is unknown. The volume did not appear on an inventory of the castle's collection until the eighteenth century.

Waldseemüller's world map of 1507 includes a tribute to Amerigo Vespucci within its border. Vespucci's portrait is given equal status with that of Ptolemy, and Vespucci is seen viewing the Western Hemisphere, which is dominated by a recognizable North America and South America connected by an isthmus. The twelve sheets that create a map measuring thirty-four square feet present a revolutionary geography that raises a series of questions about the map. Although North America and South America are, unlike on the inset, separated by a strait, the geography is surprisingly recognizable and seemingly prescient. There is a suggestion of a Florida peninsula six years before Ponce de León is supposed to have discovered it. There is a reasonable depiction of the Gulf Coast, the Yucatán Peninsula, and the Gulf of Mexico twelve years before Pineda drew the first known chart of the region. And, most pertinent to this saga, the southern continent in the New World is named "America."

The discovery of the world map of 1507 uncovered the source for the Stobnicza hemispheric map of 1512, which replicated the insets on the Waldseemüller map. More significantly, the 1507 world map replaced the 1520 map by Peter Apian[us] in earning the designation as "the first map to name America."

The discovery also initiated an immediate frenzied attempt to purchase the map and bring it to the nation that it named.

The initial probing came from the most notable procurer of rare Americana at the time, Henry N. Stevens. He had built the collections of libraries and individuals, but his name would reappear related to an important controversy concerning the Waldseemüller world map of 1507. At first he tried to purchase the map for the John Carter Brown Library of Americana in Providence, Rhode Island. Shortly thereafter, in 1910, at the time that he offered to sell facsimiles of both the world map of 1507 and the 1516 "Carta Marina," Stevens announced that he was serving as the agent for sale of the original world map of 1507. The price was set at $300,000. That offer was rapidly withdrawn, and over the ensuing years the names of J. Pierpont Morgan, Henry E. Huntington, and Paul Mellon appear on the list of interested parties.

The first half of the twentieth century was marked by sporadic attempts at purchase or sale, but, in general, the reigning count or prince of Wolfegg Castle was not interested in parting with the volume, which remained sequestered in an ordinary large standing safe on the second floor of the castle. At the end of World War II, there was immediate concern regarding the status of the map, and inquiries were made. While the university buildings in Freiburg im Breisgau, where Waldseemüller studied, were totally destroyed, Wolfegg Castle and its treasure were unharmed.

Beginning in October 1983 and extending through February 1986, as a gesture of friendship from Count Willibald Waldburg-Wolfegg, the map was placed on loan to the Smithsonian Institution's National Museum of American History. Housed in a hermetically sealed transparent case, the volume was opened to the sheet that presented South America with the word "America." At the time of preparation for the exhibit, the document was investigated by a printing expert at the museum.

Analysis of the Waldseemüller world map of 1507 demonstrated that the imprint was made by woodblocks and font, both of which demonstrated signs of aging when compared with the 1516 "Carta

PLATE 15. Sheet 1 (top row, left), Waldseemüller world map of 1507.

PLATE 16. Sheet 5 (middle row, left), Waldseemüller world map of 1507.

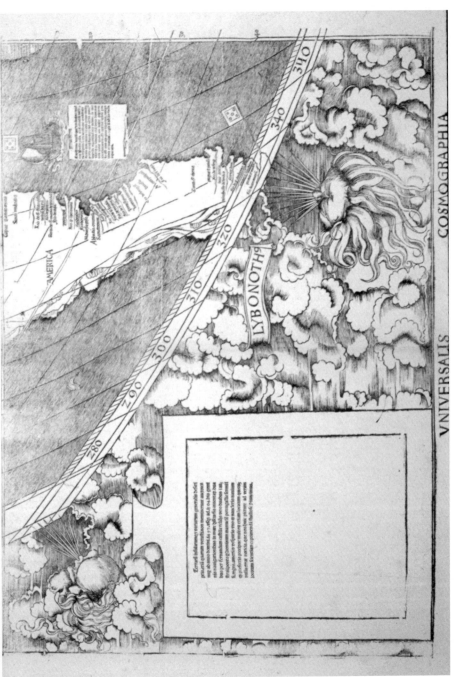

PLATE 17. Sheet 9 (bottom row, left), Waldseemüller world map of 1507.

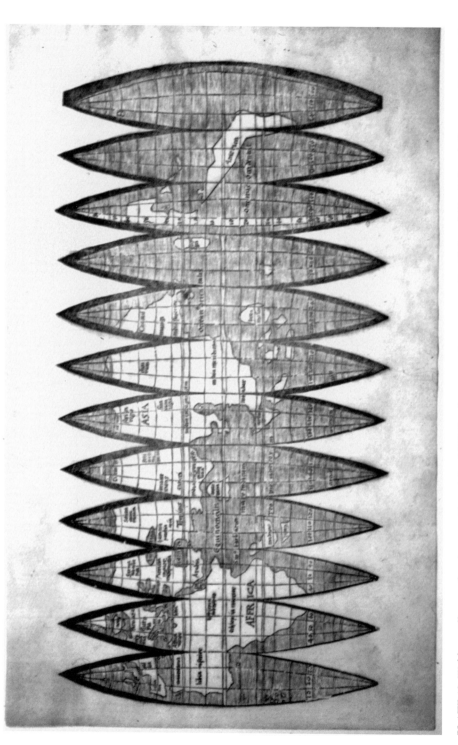

PLATE 18. Waldseemüller set of twelve gores for a globe, 1507, Strasbourg, woodcut on paper, 7 x 13.5 in. (180 x 345 mm). Courtesy of James Ford Bell Collection, University of Minnesota, St. Paul, Minnesota.

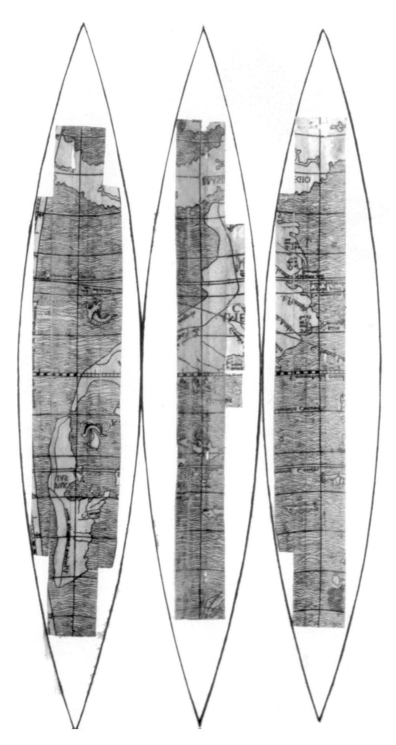

PLATE 19. Johann Schöner strips from gores for terrestrial globe, 1515, Nuremburg, woodcut on parchment. Courtesy of Jay Kislak Collection, Library of Congress, Washington, DC.

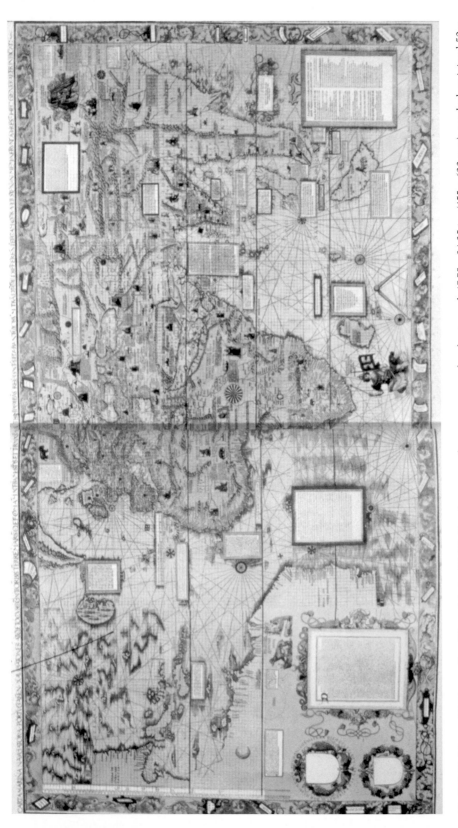

PLATE 20. Martin Waldseemüller, "Carta Marina," 1516, Strasbourg, woodcut on paper, twelve sheets, each 17.75 x 24.25 in. (455 x 620 mm), total when joined 50 x 95 in. (1,320 x 2,360 mm).

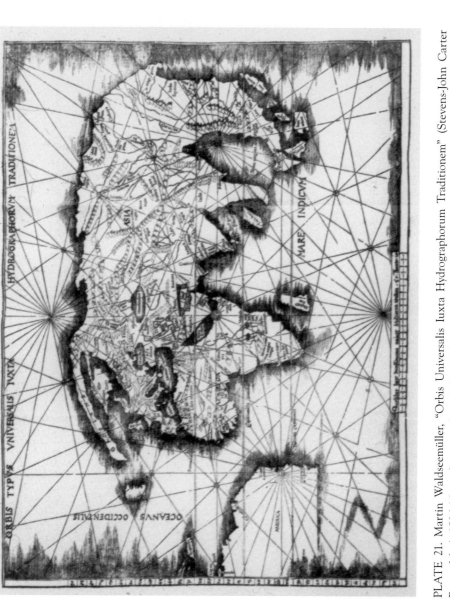

PLATE 21. Martin Waldseemüller, "Orbis Universalis Iuxta Hydrographorum Traditionem" (Stevens-John Carter Brown Map), 1506, Nuremberg, woodcut, 17.5 x 22.5 in. (445 x 570 mm). Insert in 1513 Ptolemy, *Geographiae.* Courtesy of John Carter Brown Library, Providence, Rhode Island.

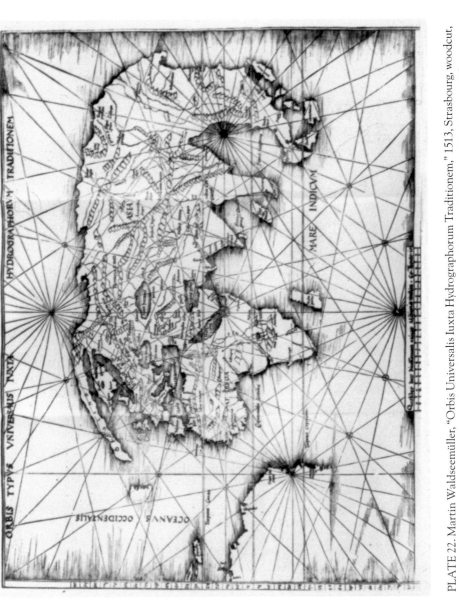

PLATE 22. Martin Waldseemüller, "Orbis Universalis Iuxta Hydrographorum Traditionem," 1513, Strasbourg, woodcut, 17.5 x 22.5 in. (445 x 570 mm). From Ptolemy, *Geographiae*. Courtesy of Library of Congress, Washington, DC.

Marina." Also, it was noted that the verso (back side) of the sheet that had been pasted onto the blank rectangle in the right lower portion of the Waldseemüller map consisted of text from a book that was printed in 1515. Therefore, it was concluded that the copy of the Waldseemüller in hand was probably produced at the same time as the "Carta Marina"—in 1516 rather than 1507.

Seven years later, in 1992, the Waldseemüller world map returned to Washington, DC, as part of the celebration of the five-hundredth anniversary of Columbus's discovery of land in the Western Hemisphere. On that occasion, the map was hung with its twelve sheets appropriately juxtaposed, constituting the first time that the map was displayed in its entirety.

At that time, the possibility that the map could be purchased by the Library of Congress was raised. Over the ensuing nine years, negotiations were prolonged and complicated by the difficulty in procuring the necessary export license from the German government because the map was on its list of national treasures. Chancellor Gerhard Schröder's defeat of Helmut Kohl in the national elections of 1998 was a critical element in the consummation of the sale from Prince Waldburg-Wolfegg-Waldsee to the Library of Congress.

The purchase of Waldseemüller's world map of 1507 constitutes the largest expenditure on the part of the Library of Congress for a single item. It was unveiled at that facility on July 23, 2003, during the opening of a Lewis and Clark exhibition. The world map of 1507 will be given an appropriate position of prominence in the Pavilion of the Discoverers that is reached by passing through the gallery that displays "America's Treasures" in the Thomas Jefferson Building of the Library of Congress, the world's most complete repository of maps.

As a glorious addendum, through the generosity of the Jay Kislak Foundation, the 1507 world map has been joined by the maps that shared its binding at Wolfegg Castle. The Library of Congress displays the 1516 "Carta Marina," the Schöner celestial gores, and the fragments of the terrestrial gores as part of the Kislak Collection in an area adjacent to the rare book room.

Questions abound regarding the document that put "America" on the map.

Is there an explanation for the seemingly anticipatory geography of the New World as depicted on the map? Does the representation of the two continents in the Western Hemisphere—interposed between two great oceans before the Pacific Ocean was viewed from continental land and before Magellan's ships passed through the strait bearing the explorer's name—suggest previous explorations that were the subjects of lost chronicles? What was the basis for the Florida and Yucatán peninsulas and the Gulf of Mexico, as they are intimated on the map, in view of the fact that there is no evidence of their discovery until after the map was printed? Is the cartographic depiction simply a matter of chance or the musings of the mapmaker?

Along etymological lines, what was the source for the word "America"? Was it Amerigo Vespucci or Richard Amerike, a Bristol customs official? If it was Vespucci, as most have concluded, was it appropriate, and will the many aspects of the mystery of assigning that name ever be unraveled?

Will the typographic analysis providing convincing evidence that the map, purchased by the Library of Congress, was actually printed sometime after 1515 be refuted? Will another copy ever be uncovered? The possibility is made more feasible by the finding of an identical offset image of the back of the sheet on which the name "America" appears.

Which truly is the earliest map with the name "America"? Is it the Waldseemüller world map of 1507, or is it the map that resides in the John Carter Brown Library? Will the issue of priority ever be resolved, and does it affect the designation of the map in the Library of Congress as America's Birth Certificate or Baptismal Document?

Somewhat surprisingly, perhaps related to the recent publicity, the world map of 1507 has found its way into modern American poetry and prose. In 1998, a colorful dust jacket adapted from Waldseemüller's world map appeared as a cover for a book consisting of new and selected poems by Lucia Perillo. The book, *The Oldest Map with the Name America*, takes its title from a poem in which the map is used as a springboard for the poet's reflections. The poem juxtaposes considerations regarding the map with introspective thoughts of the poet.

Within Lucia Perillo's four stanzas of blank verse, in which the Waldseemüller world map of 1507 is referenced, the reader is informed that the featured map and its companion, the 1516 "Carta Marina," lay aging for centuries at Wolfegg Castle. The cartographic characteristics specified by the poet include the map's "heart" shape and the representation of the continent as a "geometric" thin strip and with an empty interior. Waldseemüller is credited for avoiding the practice of his contemporary cartographers—the inclusion of ornamental "pseudohuman freaks" and "rumored continents." The poet's description of the 1516 "Carta Marina" emphasizes the replacement of the name "America" by "Terra Cannibalor," the inclusion of scenes of cannibalism, and the fact that the continent had grown to such an extent that there was no room on the map to include a west coast.[1]

The 2003 recipient of both the Pulitzer Prize for Literature and the National Book Critics Circle Award, *The Known World* by Edward P. Jones not only contains a passage in which the Waldseemüller map is described, but, according to the author, the map contributed to the selection of the title because a legend on the map referred to "the Known World." The map appears, by name, in the description of the office of the white sheriff of an imaginary Virginia county in the pre–Civil War South in this novel about black slave owners.

According to the novel, in the spring of 1844, in the jail of Manchester County, Virginia, a conversation took place between John Skiffington, the sheriff, and his only prisoner, a Frenchman who murdered his partner. The map hanging on the wall is described as "a browned and yellowed woodcut of some eight feet by six feet. The map had been created by a German, Hans Waldseemuller, who lived in France three centuries before, according to a legend in the bottom right-hand corner." Apparently, a Russian who claimed to be a descendant of the man who made the map had passed through the

Virginia town, and the sheriff bought the map from him. "The map came from the Russian in twelve parts, each weighing about three pounds." According to the sheriff, the Russian told him "that it was the first time the word *America* was put on a map." "The land of North America on the map was smaller than it was in actuality, and where Florida should have been, there was nothing. South America seemed the right size, but it alone of the two continents was called 'America.' North America went nameless."[2]

In response to my questioning the author as to why he included that specific map, which remained undiscovered at the time that the novel took place, Edward P. Jones indicated:

> I didn't know the Waldseemuller map wasn't discovered until 1901, or I would have used a different name. I had [read] a few articles from *The Washington Post* and the *New York Times* about the map being acquired by the Library of Congress, and I don't recall either paper mentioned the 1901 date [of discovery]. (A *Post* reporter said the first name wasn't Hans, but I got that name from somewhere. It must have been from the *Times*. I wouldn't have made it up.) My original intent was to title the book "All Aunt Hagar's Children," but after those articles came along, I decided to make use of the map to say something about the world I was creating. It was a strange topsy-turvy world and I wanted something to reflect that—I, of course, added things like Florida being missing and North America. The articles didn't say there was a legend with the map so I created my own. . . . I also had to make up what the map was made of. The articles weren't very clear about that either.

The references to the Waldseemüller world map of 1507 in modern poetry and prose bring back into focus the issue of print as a factor in the dissemination of facts and fancies. The print that created a large and elegant map, which depicted a revolutionary geographic concept of a New World, also dramatically placed a new name, "America," on land in that New World. The print that allowed for the dissemination of the newly discovered geography and the momentous assignment of a place name on continental land in that new part of the world had a major impact.

The Waldseemüller world map of 1507 represents the graphic assignment of a name to an area of land. That name, which could not be erased even by the mapmaker who assigned it, was adopted, over two hundred and fifty years after its introduction, at the time thirteen colonies joined to become the United States of America. The map is associated with an improbable genesis, a mysterious interval of absence, a dramatic and serendipitous discovery, and a complex process of acquisition. The sequence of these events infuses the map cumulatively with an aura of intrigue. The map can be regarded as having the star role in a historic drama.

The Waldseemüller world map of 1507 was the largest map to be printed up to its time. It is an artistic document that depicts a recognizable anticipatory geography of the New World and was seminal in the naming of the nation in which it now resides. It has taken its place in the Thomas Jefferson Building, an edifice of unrivaled beauty, at the Library of Congress, where it is regarded as the crowning jewel of a radiant collection. The map joins the Declaration of Independence, the Constitution, and the Bill of Rights in defining the history of a nation and, as such, earns recognition as America's Birth Certificate and as the most important graphic document in the history of the United States.

MUNDUS NOVUS
NEW WORLD

Amerigo Vespucci to
Lorenzo di Pierfrancesco de' Medici,
with many salutations.

I n the past I have written to you in rather ample detail about my
return from those new regions which we searched for and discovered with the fleet, at the expense and orders of His Most Serene
Highness the King of Portugal, and which can be called a new
world, since our ancestors had no knowledge of them and they are
entirely new matter to those who hear about them. Indeed, it surpasses the opinion of our ancient authorities, since most of them
assert that there is no continent south of the equator, but merely
that sea which they call the Atlantic; furthermore, if any of them
did affirm that a continent was there, they gave many arguments to
deny that it was habitable land. But this last voyage of mine has
demonstrated that this opinion of theirs is false and contradicts all
truth, since I have discovered a continent in those southern regions
that is inhabited by more numerous peoples and animals than in our
Europe, or Asia or Africa, and in addition I found a more temperate
and pleasant climate than in any other region known to us, as you
will learn from what follows, where we shall briefly write only of the
main points of the matter, and of those things more worthy of note
and record, which I either saw or heard in this new world, as will be
evident below.

We set out from Lisbon under favorable conditions on 14 May 1501 by order of the aforesaid king, with three ships, to go in quest of new regions to the south, and we sailed steadily for twenty months, and the route was as follows. We sailed to what were formerly called the Fortunate Islands and are now the Grand Canary Islands, which are in the third climate and at the bounds of the inhabited West. From there, we traveled on the Ocean Sea along the entire African coast and part of the Ethiopian, as far as the Ethiopian promontory, as Ptolemy called it, which is now called Cape Verde by our people, and Bezeguiche by the Ethiopians. The region is Mandanga, fourteen degrees north of the equator within the Torrid Zone, and it is inhabited by black tribes and peoples. There, once we had recovered our strength and procured all the necessities for our voyage, we weighed anchor and spread our sails to the winds; and set our course across the vast Ocean toward the Antarctic, steering somewhat to the west with the wind known as Vulturnus: and from the day we left the aforesaid promontory, we sailed for two months and three days before sighting any land. What we suffered in that vast expanse of sea, what dangers of shipwreck, what physical discomforts we endured, what anxieties beset our spirits, I leave to the understanding of those who have learned well and from much experience what it means to quest after uncertain things, things they have dared to investigate without prior knowledge of them. And that I might condense the whole story into one sentence, know that out of the sixty-seven days we sailed, we had forty-four continuous days of rain, thunder, and lightening, so dark that we never saw sunlight in the day, nor clear sky at night. Fear so overwhelmed us that we had almost abandoned all hope of survival. However, in those frequent, terrible tempests of sea and sky, it pleased the Most High to show us a nearby continent, and new regions and an unknown world. Sighting them, we were filled with joy, which, as one can well imagine, seized those who have found safety after calamities and misfortunes. Thus, on 7 August 1501, we dropped anchor off the shores of those regions, thanking our God with solemn prayer and the singing of a Mass. There we learned that the land was not an island but a continent, both because it extends over very long, straight shorelines, and because it is filled with countless inhabitants. For in it we encountered innumerable peoples and tribes, and all kinds of sylvan animals not found in our regions, and many other things we had never seen before, which would take too long to describe individually. God's mercy shone about greatly when we entered those regions; for our firewood and water supplies were dwindling, and in a few days we might have perished at sea. Honor be to Him, and glory, and thanks.

We decided to sail along the shore of that continent to the east, and never to lose sight of it. Soon we came to a bend where the shore curved to the south: the distance from where we first touched land to this bend was about three hundred leagues. In this phase of the voyage we landed on several occasions and conversed in friendly fashion with the people, as you will hear below. I had forgot to write to you that from the promontory of Cape Verde to the start of that continent is a distance of about seven hundred leagues, although I estimate that we sailed more than eighteen hundred, owing in part to our ignorance of the place and the ignorance of the pilot, and in part because of the storms and winds which blocked our direct course and forced us to make frequent turns. For if my companions had not relied upon me and my knowledge of cosmography, there would have been no pilot or captain on the voyage to know within five hundred leagues where we were. Indeed, we were wandering with uncertainty, with only the instruments to show us accurate altitudes of the heavenly bodies: those instruments being the quadrant and astrolabe, as everyone knows. After this, everyone held me in great honor. For I truly showed them that, without any knowledge of sea charts, I was still more expert in the science of navigation than all the pilots in the world: for they know nothing of any places beyond those where they have often sailed before. In any case, where the aforementioned bend in the land curved southward on the coast, we agreed to sail beyond it and explore what was in those regions. Therefore we sailed along the shore, approximately six hundred leagues, and we often landed and conversed with the inhabitants of those regions, and were warmly received by them, and sometimes stayed with them fifteen or twenty days at a time, always in a very friendly and hospitable way, as you will hear in the following. Part of that new continent lies in the Torrid Zone beyond the equator and toward the Antarctic

Pole; it starts eight degrees beyond the equator. We sailed along the shore until we passed the Tropic of Capricorn and found the Antarctic Pole, fifty degrees above their horizon, and we were 17 degrees from the Antarctic Circle itself; I shall now relate in due order what we saw there and what we learned of those people's nature, customs, and tractability, and of the fertility of the soil, salubriousness of climate, the dispositions of the heavens and the heavenly bodies, and in particular of the fixed stars of the eighth sphere, which our ancestors saw or described.

First, then, the people. We found such a great multitude of people in those regions that no one could count their number (as one reads in the book of the *Apocalypse*); a gentle, tractable people. Everyone of both sexes goes about naked, covering no part of the body, and just as they issued from their mothers' wombs so they go about until their dying day. They have big, solid, well-formed and well-proportioned bodies, and their complexions tend toward red, which happens, I suppose, because in going about naked they are colored by the sun. They also have long black hair. They are nimble in gait and in their games, and have open, pleasant faces, which they themselves, however, disfigure. They pierce their own cheeks, lips, noses, and ears, and you must not imagine that these holes are small or that they have but one of them: indeed I saw several people who had seven holes in a single face, each big enough to hold a plum. They fill these holes with beautiful stones, cerulean, marble-like, crystalline, or alabaster, or with very white bones and other things artfully wrought in their fashion; if you were to see such an unusual and monstrous thing as a man with seven stones just in his cheeks or jaws or lips, some of them half a palm long, you would be amazed. And I often considered this and judged that seven such stones must weight sixteen ounces. Beyond that, in each ear, which they pierce with three holes, they carry more stones dangling from rings; this custom is only for the men: the women do not pierce their faces, but only their ears. They have another custom that is appalling and passes belief. Their women, being very lustful, make their husbands' members swell to such thickness that they look ugly and misshapen; this they accomplish with a certain device they have and by bits from certain poisonous animals. Because of this, many men lose their members, which rot through neglect, and they are left eunuchs. They have no cloth of wool, linen, or cotton, since they need none. Nor have they private property, but own everything in common: they live together without a king and without authorities, each man his own master. They take as many wives as they wish, and soon may couple with mother, brother with sister, cousin with cousin, and in general as often as they please, observing no order in any of these matters. Moreover, they have no temple and no religion, nor do they worship idols. What more can I say? They lie according to nature, and might be called Epicureans rather than Stoics. There are no merchants among them, nor is there any commerce. The peoples make war among themselves without art or order. The elders deliver orations to the young to sway their will, urging them on to wars in which they kill each other cruelly, and they take captives and keep them, not to spare them, but to kill them for food: for they eat each other, the victors eat the vanquished, and together with other kinds of meat, human flesh is common fare among them. This you may be sure of, because one father was known to have eaten his children and wife, and I myself met and spoke with a man who was said to have eaten more than three hundred human bodies; and I also stayed twenty-seven days in a certain city in which I saw salted human flesh hanging from house-beams, much as we hang up bacon and pork. I will say more: they marvel that we do not eat our enemies and use their flesh as food, for they say human flesh is very savory. Their weapons are bows and arrows, and when they charge into battle, they cover no part of their bodies to protect themselves, also in this respect like animals. We tried our best to dissuade them from these wicked customs, and they promised us that they would give them up. Their women, as I said, although they go naked and are exceedingly lustful, still have rather shapely and clean bodies, and are not as revolting as one might think, because, being fleshy, their shameful parts are less visible, covered for the most part by the good quality of their bodily composition. It seemed remarkable to us that none of them appeared to have sagging breasts, and also, those who had borne children could not be distinguished from the virgins by the shape or tautness of their wombs, and this was true too of other parts of their bodies, which decency bids me pass over. When they were able to copulate with Christians, they were driven by their excessive lust to corrupt and prostitute all their modesty. The people live to be 150 years old, seldom fall ill, and if they do happen to contract some sickness, they cure themselves with certain roots of herbs. These are the more remarkable things I noticed among them. The air there is very temperate and good, and, as I was able to learn by conversing with the people, there is no pestilence or illness there deriving from contaminated air, and unless they die a violent death, they live a long life: I think this is due to the southern winds blowing constantly there, especially the one we call Eurus, which is to them as Aquilo is to us. They are very zealous fishermen, and the sea there is full of fish of all sorts. They are not hunters: I think this is because there are many kinds of forest animals there, especially lions, bears, countless snakes, and other dreadful and ill-formed beasts, and forests on all sides with trees of enormous size, that they do not dare to expose themselves, naked and without any protection or weapons, to such dangers.

The land of those regions is very fertile and pleasant, abundant in hills and mountains, countless valleys and huge rivers, watered by healthful springs, and filled with broad, dense, barely penetrable forests and all sorts of wild beasts. Great trees grow there without cultivation, and many of them produce fruits delicious to taste and beneficial to the human body, though several indeed are the opposite, and none of the fruits there are like our own. Numberless kinds of herbs and roots grow there as well, from which the people make bread and excellent foods. They also have many seeds, totally different from ours. There are no kinds of metal there except gold, in which those regions abound, although we did not bring any back with us on this our first voyage. The inhabitants apprized us of it, and told us that in the interior there is great abundance of gold, which they do not at all value or consider precious. They are rich in pearls, as I wrote to you elsewhere. If I wanted to mention separately all the things which are there, and to write about the numerous kinds of animals and their great numbers, I would grow too prolix with a matter so vast; and I certainly believe that our Pliny did not come within a thousandth part of the types of parrots and other birds and animals which are in those regions, with such great diversity of forms and colors that even Polycletus, master of painting in all its perfection, would have failed to depict them adequately. All the trees there are fragrant, and all produce gum or oil or some liquor, and I do not doubt that their properties, if they were known to us, would be salubrious for the human body; and certainly, if anywhere in the world there exists an Earthly Paradise, I think it is not far from those regions, which lie, as I said, to the south, and in such a temperate climate that they never have either icy winters or scorching summers.

Sky and air are clear for most of the year and free from dense vapors. The rains there fall delicately and last three or four hours, then vanish like mist. The sky is adorned with very beautiful signs and figures, in which I noticed twenty stars as bright as we sometimes see Venus or Jupiter. I considered their movements and orbits and measured their circumferences and diameters with geometric methods, and determined that they are of great magnitude. I saw three Canopi in that sky, two of which are bright indeed, and the third dim. The Antarctic Pole has no Ursa Major and Ursa Minor, as appear in our Arctic Pole, nor is any bright star seen near it; of the stars which are carried around it in smaller orbit, there are three which form the figure of an orthogonal triangle, of which half the circumference, or the diameter, is 9 degrees. As they rise, a white Canopus of extraordinary size can be seen to the left. When they reach mid-heaven they form this figure:

After these come two other stars, of which half the circumference, or diameter, is 12 degrees, and with them can be seen another white Canopus. Another six stars, the brightest and most beautiful of all in the eighth sphere, follow them; on the surface of the firmament, these stars have a half-circumference, diameter, of thirty-two degrees. A black Canopus of immense size soars up with them. They are seen in the Milky Way and form a figure like this when they are on the meridian line:

I encountered many other beautiful stars during this voyage of mine, and I notated their movements carefully and have described them beautifully and graphically in a booklet of mine. His Most Serene Highness the King has it at present, and I hope that he will return it to me. In that hemisphere I saw things which do not agree with the arguments of philosophers: a white rainbow was seen twice around midnight, not only by me, but also by all the sailors. Likewise, several times we saw a new moon on the day when it was in conjunction with the sun. Every night in that part of the sky, innumerable vapors and bright flares streak across. A bit earlier I spoke of the hemisphere, although, properly speaking, it is not fully a hemisphere with respect to ours; but since it approaches the shape of one, it is permissible to call it so.

Therefore, as I said, from Lisbon, our point of departure, 39 degrees from the equator, we sailed fifty degrees beyond the equator, which together make about ninety degrees, and since this sum makes a quarter of the great circle, according to the true reasoning of measurement passed on to us by the ancients, it is clear that we sailed around a quarter of the world. And by this logic, we who live in Lisbon, 39 degrees this side of the equator in the northern latitude, are at an angle of five degrees in the transverse line to those who live at the fiftieth degree beyond the same line in the southern latitude, or, so that you may understand more clearly: a perpendicular line, which hangs over our heads from a point directly above us while we stand upright, hangs pointing toward their sides or ribs: thus we are in an upright line, and they in a transverse line, and a kind of orthogonal triangle is formed thereby, of which we form the perpendicular line, they the base, and the hypothenuse extends from our vertex to theirs, as is evident in the drawing. And let these words suffice for cosmography.

These were the more noteworthy things I saw on this last nav-igation of mine, which I call the "third journey." The other two "journeys" were my two other navigations, which I made toward the west on a mandate from the Most Serene King of the Spains; on those voyages I noted the marvels accomplished by the sublime cre-ator of all, our God: I kept a diary of the noteworthy things, so that, if ever I am granted the leisure, I may gather together all these mar-vels one by one and write a book, either of geography or of cosmog-raphy, so that my memory will live on for posterity, and so that the immense creation of almighty God, unknown in part to the ancients yet known to us, may be recognized. I pray, therefore, to the most merciful God that He may prolong the days of my life, that by His good grace, and for the salvation of my soul, I may attain the fullest realization of my goals. The other two "journeys" I keep among my private papers, and when the Most Serene Highness returns the "third journey" to me, I shall try to return to tranquillity and my homeland, where I will be able to confer with experts and, with help and encouragement of my friends, to complete that work.

I ask your forgiveness for not sending you this last navigation, or rather this last "journey," as I had promised to do in my last letter: you know the reason, since I could not yet have the original back from His Most Serene Highness. I still plan to make a fourth voyage, and have already received the promise of two ships together with their equipment, so that I may prepare to search for new regions to the south, travelling from the east with the wind called Africus; on this voyage I think I will accomplish many things to the praise of God, the benefit of this kingdom, and the honor of my old age; and I await nothing but the consent of His Most Serene Highness. May God permit whatever is for the best. You will learn of whatever happens.

The interpreter Giocondo has translated this letter from Italian to Latin, so that all the Latins may understand how many marvelous things are being discovered every day, and to curb the audacity of those people who wish to study the heavens and their majesty and to know more than they are permitted to know, for, ever since the world began, the earth's vastness and all things contained in it have been unknown.

Praise to God.

LETTERA: QUATTOR AMERICI VESPUCIJ NAUIGATIONES

The Four Voyages of Amerigo Vespucci

To the most illustrious René, King of Jerusalem and of Sicily, Duke of Lorraine and Bar, Amerigo Vespucci pays humble homage and presents appropriate recommendations.

Perchance, most illustrious King, your majesty will be astonished at my foolhardiness, because I feel no apprehension in addressing to you the present long letter, even though I know you to be incessantly occupied with matters of the highest importance and with numerous affairs of State. And I shall be considered not only a presumptuous man but one who has accompanied a useless work in undertaking to send you also a story which hardly concerns your position, addressed by name to Ferdinand, King of Castile, and written in an unattractive and quite unpolished style, as if I were a man unacquainted with the Muses and a stranger to the refining influence of learning. My trust in your merits, and the absolute truth of the following accounts (on matters which neither ancient nor modern authors have written), will perhaps excuse me to your Majesty.

I was urged to write chiefly by the bearer of the present letters, Benvenuto, a humble servant of your Majesty and a friend of whom I need not be ashamed. When this gentleman found me at Lisbon, he begged me to acquaint your Majesty with the things seen by me during my four voyages to different quarters of the globe. For, you must know that I have completed four voyages of discovery to new lands: two of them were undertaken by the order of Ferdinand, the illustrious King of Castile, and carried me toward the West, through the Great Gulf of the Ocean; the other two were undertaken at the command of Manuel, King of Portugal, and carried me toward the south.

I have therefore prepared myself for the task urged upon me by Benvenuto, hoping that your Majesty will not exclude me from the number of your insignificant servants, especially if you recollect that formerly we were good friends. I refer to the years of our youth, when we were fellow-students, and together drank in the elements of grammar under the holy and venerable friar of St. Mark, my uncle, Friar Giorgio Antonio Vespucci—a man of good life and tried learning. Had it been possible for me to follow in his footsteps, I should be quite a different man to-day, as Petrarch says. However that may be, I am not ashamed of being what I am; for I have always taken pleasure in virtue for its own sake and in scholarship. If, then, these narratives give you no pleasure whatever, I shall repeat the words which Pliny once wrote to Mæcenas, "Formerly you were wont to take delight in my pleasantry." Your Majesty, it is true, is ever occupied with affairs of State; still, you can secretly steal just a little time in which to read these accounts, trifling though they be. I assure you that their very novelty will please. You will find in these pages no slight relief from the wasting cares and problems of government. My book will serve you as the sweet fennel, which, when taken after meals, is wont to leave a pleasant breath and to promote better digestion.

If, by chance, I have been more prolix than the subject warrants, I crave your indulgence.

Farewell.

PREFACE

Most Illustrious King! Your Majesty must know that I came to this country primarily as a merchant. I continued in that career for the space of four years. But when I observed the various changes of fortune, and saw how vain and fleeting riches are, and how for a time they lift man to the top of the wheel and then hurl him headlong to the bottom—him, who had boasted of wide possession;—when I saw all this, and after I had personally suffered such experiences, I determined to abandon the business career and to devote all my efforts to worthier and more enduring ends.

And so I set about visiting different parts of the world and seeing its many wonders. Both time and place were favorable to my plans. For Ferdinand, King of Castile, was at that time fitting out four ships to discover new lands in the west, and His Highness made me one of that company of explorers. We set sail from the harbor of Cadiz on the 20th of May, 1497, making our way through the Great Gulf of the Ocean. This voyage lasted eighteen months, during which we discovered many lands and almost countless islands (inhabited as a general rule), of which our forefathers make absolutely no mention. I conclude from this that the ancients had no knowledge of their existence. I may be mistaken; but I remember reading somewhere that they believed the sea to be free and uninhabited. Our poet Dante himself was of this opinion, when, in the 18th canto of the Inferno, he pictures the death of Ulysses. From the following pages, however, your Majesty will learn of the marvels I saw.

A description of the chief lands and of various islands, of which ancient authors make no mention, but which recently, in the 1497th year from the incarnation of Our Lord, were discovered in the course of four ocean voyages undertaken by order of their Serene Highnesses of Spain and Portugal. Of these voyages, two were through the western sea, by order of King Ferdinand of Castile; the remaining two were through southern waters, by order of Manuel, King of Portugal. To the above mentioned Lord Ferdinand, King of Castile, Amerigo Vespucci, one of the foremost captains

and commanders of that fleet, dedicates the following account of
the new lands and islands.

THE FIRST VOYAGE

In the year of Our Lord 1497, on the 20th day of May, we set sail
from the harbor of Cadiz in four ships. On our first run, with the
wind blowing between the south and the southwest [Vespucci
names the wind according to the point toward which it blows], we
made the islands formerly called the Fortunate Islands, but now the
Grand Canary, situated at the edge of the inhabited west and within
the third climate. At this place, the North Pole rises 27 2/3 degrees
above the horizon, the islands themselves being 280 leagues from
the city of Lisbon, in which this present pamphlet was written.
There we spent almost eight days, providing ourselves with fuel and
water and other necessary things. Then, after first offering our
prayers to God, we raised and spread our sails to the wind, shaping
our course to the west, with a point to the southwest. We kept on
this course for some time, and just as the 27th day was past we
reached an unknown land, the mainland as we thought. It was distant from the islands of the Grand Canary 1000 leagues, more or
less; it was inhabited, and was situated in the Torrid Zone. This we
ascertained from the following observations: that the North Pole
rises 16 degrees above the horizon of this new land, and that it is 75
degrees more to the west than the islands of Grand Canary—at least
so all our instruments showed.

Here we dropped the bow anchors and stationed our fleet a
league and a half from the shore. We then lowered a few boats, and,
filling them with armed men, we pulled as far as the land. The
moment we approached, we rejoiced not a little to see hordes of
naked people running along the shore. Indeed, all those whom we
saw going about naked seemed also to be exceedingly astonished at
us, I suppose because they noticed that we wore clothing, and presented a different appearance from them. When they realized that
we had actually arrived, they all fled to a hill near by; and though

we beckoned to them and made signs of peace and friendship, we could not induce them to approach. When night closed rapidly upon us, we felt some fear in trusting our ships in such a dangerous roadstead, for there was here no protection against violent seas. We therefore agreed to depart early the next morning in search of some harbor where we might station our ships in a safe anchorage. After we had formed this resolution, we spread our sails to a gentle breeze blowing along the shore, keeping land always in sight and continually seeing the inhabitants along the beach. In this way we sailed for two whole days, and discovered a place quite suited to our ships, where we anchored only one-half league from the land. Here we again saw countless hordes of people. Desiring to see them close by and to speak with them, on that very day we approached the shore in our boats and skiffs, and then we landed in good order, about forty strong. The natives, however, showed themselves very loath to approach us or have anything to do with us. We could do nothing to induce them to speak with us or to enter upon any kind of communication. But finally, by dint of much labor undertaken with this one purpose in view, we managed to allure a few of them by giving them little bells and mirrors and pieces of crystal and other such trifles. In this way they became quite easy about us. They now came to meet us, and in fact to treat concerning terms of peace and friendship. At nightfall we took leave of them and returned to our ships. The next day, when the sun was quite risen, we again saw upon the beach an endless number of men and women, the latter carrying their children with them. We furthermore noticed that they were bringing with them all their household utensils, which will be described below in their proper place. The nearer we approached the shore, more and more of the natives jumped into the water (for there are many expert swimmers among them), and swam out the distance of a crossbow shot to meet us. They received us kindly, and in fact mingled among us with as complete assurance as if we had often met before and had frequently had dealings together. At this we were then very little pleased.

And now (so far as occasion permits), we shall devote some space to a description of their customs,—such as we were able to observe.

Of the Customs of the Natives and Their Mode of Life

In regard to their life and customs, all of them, both men and women, go about entirely naked, with no more covering for their private parts than when they were born. The men are of medium size, but are very well proportioned. The color of their skin approaches red, like the hair of a lion, and I believe that, if it were their custom to wear clothing, they would be as fairskinned as we are. They have no hair on their body, with the exception of that on the head, which is long and black, particularly that of the women, who are beautiful for this very reason. Their features are not very handsome, because they have broad cheek-bones like the Tartars. They do not allow any hair to grow on their eyebrows nor their eyelids nor anywhere on their body (with the exception of the head), for this reason,—because they deem it coarse and animal-like to have hair on the body.

All of them, both men and women, are graceful in walking and swift in running. Indeed, even their women (as we have often witnessed) think nothing of running a league or two, wherein they greatly excel us Christians. They all swim remarkably well, in fact better than one would believe possible; and the women are far better swimmers than the men, a statement which I can make with authority, for we frequently saw them swim in the sea for two leagues without any assistance whatsoever.

Their weapons are the bow and arrow, which they have learned to make very skillfully. They are unacquainted with iron and the metals, and consequently, in place of iron, they tip their arrows with the teeth of animals and fishes, and they also often harden the arrows by burning their ends. They are expert archers, with the result that they strike with their arrows whatever they aim at. In some places also the women are very skillful with the bow and arrow. They have other weapons also, such as spears or stakes sharpened at the ends, and clubs with wonderfully carved heads.

They are wont to wage war upon neighbors speaking a different language, fighting most mercilessly and sparing none, except to reserve them for more cruel torture later. When they go forth to battle, they take their wives with them, not that they too may participate in the fight, but that they may carry behind the fighting men all the necessary provisions. For, as we ourselves have often seen, any women among them can place on her back, and then carry for thirty or forty leagues, a greater burden than a man (and even a strong man) can lift from the ground. They have no generals and no captain; in fact, since every one is his own leader, they go forth to war in no definite order. They never fight for power or territory, or for any other improper motive. Their one cause for war is an enmity of long standing, implanted in them from olden times. When questioned concerning the cause of such hostility, they give no other reason except that it is to avenge the death of their ancestors. Living as they do in perfect liberty, and obeying no man's word, they have neither king nor lord.

They are, however, especially inclined to war, and gird themselves for braver efforts when one of their own number is either a captive in the hands of the enemy or has been killed by them. In that case the oldest blood-relation of the prisoner or murdered man rises, rushes forth into the roads and villages, shouting and calling upon all, and urging them to hasten into battle with him to avenge the death of his kinsman. All are quickly stirred to the same feeling, gird themselves for the fight and make a sudden dash upon their enemies.

They observe no laws, and execute no justice. They do not punish their evildoers; indeed, not even the parents rebuke or chastise their children; and, wonderful to relate, we several times saw them quarrel among themselves. They are simple in their speech, but very shrewd and crafty. They speak rarely; and when they do speak, it is in a low tone, using the same sounds as we. On the whole they shape their words either on the teeth or lips, employing, of course, different words from those of our language. They have many differing idioms, for we have found such a variety of tongues in every hundred leagues that they do not understand one another.

They observe most barbarous customs in their eating; indeed, they do not take their meals at any fixed hours, but eat whenever they are so inclined, whether it be day or night. At meals they recline on the ground, and do not use either tablecloths or napkins, being entirely unacquainted with linen and other kinds of cloth. The food is served in earthen pots which they make themselves, or else in receptacles made out of half-gourds. They sleep in a species of large net made of cotton and suspended in the air; and though this mode of sleeping may appear odd and uncomfortable, I testify that, on the contrary, it is very pleasant; for it was frequently my lot to sleep in such nets, and I had a feeling of greater comfort then than when under the coverlets which we had with us.

In their person they are neat and clean, for the reason that they bathe very frequently.

In their sexual intercourse they have no legal obligations. In fact, each man has as many wives as he covets, and he can repudiate them later whenever he pleases, without its being considered an injustice or disgrace, and the women enjoy the same rights as the men. The men are not very jealous; they are, however, very sensual. The women are even more so than the men. I have deemed it best (in the name of decency) to pass over in silence their many arts to gratify their insatiable lust. They are very prolific in bearing children, and do not omit performing their usual labors and tasks during the period of pregnancy. They are delivered with very little pain,—so true is this that on the very next day they are completely recovered and move about everywhere with perfect ease. In fact, immediately after the delivery they go to some stream to wash, and then come out of the water as whole and as clean as fishes. However, they are of such a cruel nature and harbor such violent hatreds that, if the husbands chance to anger them, they immediately commit some wrong. For instance, to appease their great wrath, they kill the fetus within their own wombs, and then cause an abortion. In this way countless offspring are destroyed. They have handsome, well-proportioned and well-knit figures; indeed, no blemish can possibly be discovered in them . . .

No one of this race, as far as we saw, observed any religious law. They can not justly be called either Jews or Moors; nay, they are far worse than the gentiles themselves or the pagans, for we could not discover that they performed any sacrifices nor that they had any special places or houses of worship. Since their life is so entirely given over to pleasure, I should style it Epicurean.

They hold their habitations in common. Their dwellings are bell-shaped, and are strongly built of large trees fastened together, and covered with palm leaves, which offer ample protection against the winds and storms. In some places these dwellings were so large that we found as many as six hundred persons living in a single building. Of all these dwellings we found that eight were most thickly populated; in fact, that ten thousand souls lived within them at one and the same time. Every eight or seven years they move the seat of their abodes. When asked the reason for this, they gave a most natural answer. They said that it was on account of the continual heat of a strong sun, and because, from dwelling too long in the same place, the air became infected and contaminated, and brought about various diseases of the body. And in truth, their point seemed to us to be well taken.

Their riches consist of variegated birds' feathers, and of strings of beads (like our *pater nosters*), made of fish bones, or of green or white stones. These they wear as ornaments on the forehead, or suspended from their lips and ears. Many other such useless trifles are considered riches by them, things to which we attack no value whatever. Among them there is neither buying nor selling, nor is there an exchange of commodities, for they are quite content with what nature freely offers them. They do not value gold, nor pearls, nor gems, nor such other things as we consider precious here in Europe. In fact they almost despise them, and take no pains to acquire them. In giving, they are by nature so very generous that they never deny anything that is asked of them. But as soon as they have admitted any one to their friendship, they are just as eager to ask and to receive. The greatest and surest seal of their friendship is this: that they place at the disposal of their friends their own wives and daughters, both parents considering themselves highly honored if any one deigns to lead their daughter (even though yet a maiden) into concubinage. In this way (as I have said) they seal the bond of their friendship.

In burying the dead they follow many different customs. Some, indeed, follow the practice of inhumation, placing at the head water and food, for they believe that the dead will eat and subsist thereupon. But there is no further grief at their departure, and they perform no other ceremonies. In some places a most barbarous and inhuman rite is practiced. When any one of their fellow-tribesmen is believed to be at the point of death, his relations take him into some great forest, where they place him in one of those nets in which they are accustomed to sleep. They then suspend him thus reclining between two trees, dance around him for a whole day, and then at nightfall return to their habitations, leaving at the head of the dying man water and food to last him about four days. If at the end of this period the sick man can eat and drink, becomes convalescent, regains his health, and returns to his own habitation, then all his relations, whether by blood or marriage, welcome him with the greatest ceremonies. But there are few who can pass safely through so severe an ordeal. Indeed, no one ever visits the sick man after he is abandoned in the woods. Should he, therefore, chance to die, he receives no further burial. They have many other savage rites of burial, which I shall not mention, to avoid the charge of being too prolix.

In their sickness they employ many different kinds of medicines, so different from ours and so discordant with our ideas that we wondered not a little how any one could possibly survive. For, as we learned from frequent experience, if any one of them is sick with fever, they immerse and bathe him in very cold water just when the fever is at its height. Then they compel him to run back and forth for two hours around a very warm fire until he is fairly aglow with heat, and finally lead him off to sleep. We saw very many of them restored to health by this treatment. Very frequently they practice also dieting as one of their cures, for they can do without food and drink for three or four days. Again, they commonly draw blood, not from their arms (with the exception of the shoulder-blade), but from their loins and the calves of their legs. Often they bring about vomiting by chewing certain herbs which they use as medicine; and they have, in addition, many other cures and remedies which it would be tedious to enumerate.

They are full-blooded and phlegmatic, owing to the food they eat, which consists chiefly of roots, fruits, herbs, and fishes of different kinds. They do not raise crops of spelt or of any other grain. Their most common food is a certain root which they grind into a fairly good flour and which some of the natives call *iucha*, others *chambi*, and still others *ygnami*. [The Italian text gives *iuca*, *cazabi*, and *ignami*.] They very rarely eat flesh, with the exception of human flesh; and in this they are so inhuman and so savage as to outdo even the wild animals. Indeed, all the enemies whom they either kill or capture, without discriminating between the men and the women, are relished by them with such savageness that nothing more bar-barous and cruel can either be seen or heard of. Time and again it fell to my lot to see them engaged in this savage and brutal practice, while they expressed their wonder that we did not likewise eat our enemies. Your royal Majesty may rest assured on this point, that their numerous customs are all so barbarous that I can not describe them adequately here. Therefore, considering the many, many things I saw in my four voyages—things so entirely different from our customs and manners—I have prepared and completed a work which I have entitled "The Four Voyages." In this book I have col-lected the greater part of the things I saw, and have described them as clearly as my small ability would permit. I have not, however, published it as yet. In this work, each topic is given more careful and individual attention, and therefore in the present pamphlet I shall merely touch upon them, making only general statements. And so I return to complete the account of our first voyage, from which I have made a short digression.

In the beginning of our voyage we did not see anything of great value except a few traces of gold, and this only because they pointed out to us several proofs of its existence in the soil. I suppose we should have learned much more, had we been able to understand their language. In truth, this land is so happily situated that it could not be improved. We unanimously agreed, however, to leave it and to continue our voyage further. And so, keeping land always in sight, and tacking frequently, we visited many ports, in the mean-while entering upon communications with many different tribes of those regions. After some days we made a certain harbor in which it pleased God to deliver us from a great danger.

As soon as we entered this harbor, we discovered that their whole population, that is to say, the entire village, had houses built in the water, as at Venice. There were in all about twenty large houses, built in the shape of bells (as we have said above), and resting firmly upon strong wooden piles. In front of the doors of each house drawbridges had been erected, over which one could pass from one hut to another as if over a well-constructed road. As soon as the inhabitants of this settlement noticed us they were seized with great fear, and immediately raised the drawbridges to defend themselves against us, and hid themselves within their houses. While we were watching their actions with some degree of wonder, lo and behold about twelve of their boats (which are hollowed out of the trunk of a single tree) came over the water to meet us. The occupants of these boats looked at us and at our clothes with wonder, and rowed about us in every direction, but continued to examine us from a distance. We on our part were similarly observing them, making many signs of friendship to urge them to approach us without fear. But it was of no avail. Seeing their reluctance, we began to row in their direction. They did not await our arrival, but immediately fled to the shore, making signs to us that we should await their return, which (they signified) would be shortly. Thereupon they hurried to a near by hill, returning thence accompanied by sixteen maidens. With these they embarked in the abovementioned boats and straightway returned to us. Of the maidens, four were then placed in each one of our ships, a proceeding which, as your Majesty may well believe, astonished us not a little. Then they went back and forth among our ships with their canoes, and spoke to us in such kindly manner that we began to consider them our trusty friends. While all this was going on, behold a large crowd began to swim from their houses (already described) and to advance in our direction. Though they advanced further and further, and though they were now nearing our ships, we entertained not the slightest suspicion of their actions. At this point, however, we saw some old women standing at the doors of their houses, shouting wildly and filling the air with their cries, and tearing their hair in great distress. We now began to suspect that some great danger was threatening. Immediately the girls who had been placed on board our ships leaped into the sea. Those who were in the canoes pulled off a short distance, drew their bows and began to make a vigorous attack upon us. Moreover, those who had started from their houses and were swimming over the sea toward us, were, each one of them, carrying a lance underwater. This was sure proof of their treachery, and we began not only to defend ourselves with spirit, but also to inflict serious injuries upon them. In fact, we wrecked and sank many of the canoes, with great loss of life to their occupants,—a loss which became even greater because the natives abandoned their canoes entirely and swam to the shore. About twenty of them were killed and many more were wounded. Of ours only five were injured, all of whom were restored to health, with the help of God. We managed to capture two of the girls and three men. Later we visited the houses of settlement, and upon entering found them occupied only by two old women and a sick man. We did not set fire to the house for this reason, that we feared lest our consciences would prick us. We then returned to the ships with our five captives and put them in irons, except the girls. At night, however, both girls and one of the men very shrewdly effected their escape.

On the following day we agreed to leave that port and sail on along the coast. After a run of about eighty leagues we came to another tribe entirely different from the former in language and customs. We anchored the fleet and approached the shore in our small boats. Here we saw a crowd of about 4,000 persons on the beach. As soon as they realized that we were about to land, they no longer remained where they were, but fled to the woods and forests, abandoning on the shore everything which they had with them. Leaping upon the land, we advanced along a road leading to the forest about as far as a crossbow shot. We soon came upon many tents which had been pitched there by that tribe for the fishing season. Within them, many fires had been built for cooking their meals, and animals and fishes of various kinds were being roasted. Among other things we saw that a certain animal was being roasted which looked very much like a serpent, except for the wings which were missing. It looked so strange and so terrible that we greatly wondered at its wild appearance. Proceeding onward through their tents, we found many similar serpents, whose feet were tied and whose mouths were muzzled so that they could not open them, as is done with dogs and other wild animals that they may not bite. Their whole appearance was so savage that we, supposing them to be poisonous, did not dare approach them. They are like a young goat in size, and half as long again as an arm. Their feet are very large and heavy, and are armed with strong claws; their skin is varicolored; their mouth and face like those of a serpent. From the end of the nose to the tip of their tail they are covered (along the back) with a kind of bristle, from which we decided that they were truly serpents. And yet the above-mentioned tribe eats them. That same tribe makes bread from the fishes which they catch in the sea, the process being as follows: First of all they place the fish in the water and boil it for some time; then they pound it and crush it and make it into small cakes which they bake upon hot ashes and which they then eat. Upon tasting them we found them to be not at all bad. They have many other kinds of food, including different fruits and herbs, but it would take too long to describe them.

But to return to our story. Although the natives did not reappear from the woods to which they had fled, we did not take away any of their possessions, in order that we might increase their confidence in us. In fact, we left many small trifles in their tents, placing them where they would be seen, and at night returned to our ships. On the next day, when Titan began to rise above the horizon, we saw a countless multitude upon the shore. We immediately landed; and though the natives still appeared to be somewhat afraid of us, yet they mingled among us, and began to deal and to converse with us with complete security. They signified to us that they would be our friends, that the tents which we saw were not their real houses, and that they had come to the shore to fish. Therefore they begged us to accompany them to their villages, assuring us that they wished to welcome us as friends. We were made to understand that the cause of the friendship which they had conceived for us was our arrest of those two prisoners, who turned out to be enemies of theirs. And so, seeing the persistence with which they asked us, twenty-three of us decided to go with them, fully armed and with the firm resolve to die valiantly if need be.

After remaining there for three days, we marched inland with them for three leagues and came to a village consisting of but nine habitations. There we were received with such numerous and such barbarous ceremonies that my pen is too weak to describe them. For instance, we were welcomed with dances and with songs, with lamentations mingled with cries of joy and happiness, with much feasting and banqueting. Here we rested for the night, and the natives most generously offered us their wives. . . . After we had remained that night and half of the next day, a large and wondering crowd came to look at us, without hesitation and fear. Their elders now asked us to go with them to their villages situated farther inland, to which we again agreed. It is not an easy task to recount the honors which they showered upon us here. In short, we went about in their company for nine whole days, visiting very many of their settlements, with the result that (as we afterward learned), our companions whom we had left in the ships began to be very anxious about us and to entertain serious fears for our safety. And so, after having penetrated about eighteen leagues into the interior of the country, we decided to make our way back to the ships. On our return a great crowd of men and women met us and accompanied us all the way to the sea,—a fact which is of itself very remarkable. But there is more. Whenever it happened that one of our company would lag behind from weariness, the natives came to his assistance and carried him most zealously in those nets in which they sleep. In crossing the rivers, too (which in their country are very numerous and very large), they were so careful with the contrivances they employed that we never feared the slightest danger. Moreover, many of them, laden down with their gifts, which they carried in those same nets, accompanied us. The gifts consisted of feathers of very great value, of many bows and arrows, and of numberless parrots of different colors. Many others, also, were bringing their household goods and their animals. In fine, they all reckoned themselves fortunate if, in crossing a stream, they could bear us on their shoulders or on their backs.

However, we hastened to the sea as quickly as possible. As we were about to embark in our boats, so great was the crowding of the natives in their attempt to accompany us still further and to embark with us and visit our ships, that our boats were almost swamped by the load. We took on board, however, as many as we could accommodate and brought them to our ships. In addition to those whom we had on board, so many of them accompanied us by swimming that we were somewhat troubled by their approach; for, about a thousand of them boarded our ships (naked and unarmed though they were), and examined with wonder our equipment and arrangements and the great size of the ships themselves. And then a laughable thing happened. We desired to shoot off some of our war engines and artillery, and therefore put a match to the guns. These went off with such a loud report that a large portion of the natives, upon hearing this new thunder, leaped into the water and swam away, like frogs sitting on the bank, which jump into the bottom of the marsh and hide the moment they are startled by a noise. In this way acted the natives. Those natives who had fled to another portion of the ship were so thoroughly frightened that we repented and chid ourselves for what we had done. But we quickly reassured them, and did not permit them to remain any longer in ignorance, explaining that it was with these guns that we killed our enemies.

After entertaining them the whole day upon our ships, we warned them to depart because we intended to sail during the night; whereupon they took leave of us in a most friendly and kindly manner. We saw and learned very many customs of this tribe and region, but it is not my intention to dwell upon them here. Your Majesty will be in a position to learn later of all the more wonderful and noteworthy things I saw in each of my voyages; for I have collected them in one work written after the manner of a geographical treatise and entitled "The Four Voyages." In this work I give individual and detailed descriptions, but I have not yet offered it to the public because I must still revise it and verify my statements.

That land is very thickly populated, and everywhere filled with many different animals, very unlike those of our country. In common with us they have lions, bears, stags, pigs, goats, and fallow deer, which are, however, distinguished from ours by certain differences. They are entirely unacquainted with horses, mules, asses, dogs, and all kinds of small cattle (such as sheep and the like), and cows and oxen. They have, however, many species of animals which it would be difficult to name, all of them wild and of no use to them in their domestic affairs. But why say more? The land is very rich in birds, which are so numerous and so large, and have plumes of such different kinds and colors, that to see and describe them fills us with

wonder. The climate, moreover, is very temperate and the land fertile, full of immense forests and groves, which are always green, for
the leaves never fall. The fruits are countless and entirely different
from ours. The land itself is situated in the torrid zone, on the edge
of the second climate, precisely on the parallel which marks the
tropic of Cancer, where the Pole rises twenty-three degrees above
the horizon. During this voyage many came to look at us, marveling
at the whiteness of our skin. And when they asked us whence we
came, we answered that we had descended from heaven to pay the
earth a visit, a statement which was believed on all sides. We established in this land many baptismal fonts or baptisteries, in which
they made us baptize countless numbers, calling us in their own
tongue "charaibi"—that is to say, "men of great wisdom." The
country itself is called by them "Parias."

Later we left that harbor and land, sailing along shore and
keeping land always in view. We sailed for 870 leagues, making
many tacks and treating and dealing with numerous tribes. In many
places we obtained gold, but not in great quantities; for it sufficed us
for the present to discover those lands and to know that there was
gold therein. And since by that time we had already been thirteen
months on our voyage, and since the tackle and rigging were very
much the worse for wear and the men were reduced by fatigue, we
unanimously agreed to repair our small boats (which were leaking at
every point) and to return to Spain. Just as we had reached this conclusion, we neared and entered the finest harbor in the world. Here
we again met a countless multitude, who received us in a very
friendly manner. On the beach we built a new boat with material
taken from the other ships and from barrels and casks, placed upon
dry land our rigging and military engines, which were almost rotting
away in the water, lightened our ships and drew them up on land.
Then we repaired them and patched them, and gave them a thor-
ough overhauling. During all these occupations the inhabitants of
the country gave us no slight assistance. Indeed, they offered us provisions out of friendship and unasked, so that we consumed very
little of our own supplies. This we considered a great boon, for our
supplies at this stage were rather too meager to enable us to reach
Spain without stinting ourselves.

We remained in that port thirty-seven days, frequently visiting the villages in company with the natives and being treated with great respect by each and every one of them. When we at last expressed our intention to leave that harbor and to resume our voyage, the natives complained to us that there was a certain savage and hostile tribe, which, at a certain time of the year, came over the sea to their land, and either through treachery or through violence killed and devoured a great number of them. They added that others were led off as prisoners to the enemy's country and home, and that they could not defend themselves against these enemies, making us understand that that tribe inhabited an island about one hundred leagues out at sea. They related their story to us in such plaintive tones that we took pity on them and believed them, promising that we should exact punishment for the injuries inflicted upon them. Whereat they greatly rejoiced and of their own accord offered to accompany us. We refused for several reasons, agreeing to take seven with us on the following condition: that at the close of the expedition they should return to their country alone and in their own canoes, for we did not by any means intend to take the trouble of bringing them back. To this condition they gladly assented, and so we took leave of the natives, who had become our dear friends, and departed.

We sailed about in our refitted ships for seven days, with the wind blowing between the northeast and east. At the end of this period we reached many islands, of which some were inhabited and others not. We thereupon approached one of them; and while endeavoring to anchor our ships we saw a great horde of people on the island, which the inhabitants call Ity. After examining them for some time, we manned the small boats with brave men and three guns, and rowed nearer the shore, which was filled with 400 men and very many women, all of whom (like the others) went about naked. The men were well built, and seemed very warlike and brave, for they were all equipped with their usual arms, namely, the bow and arrow and the lance. Very many of them, moreover, bore round shields or even square shields, with which they defended themselves so skillfully that they were not hindered thereby in shooting their arrows.

When we had come in our boats to within a bowshot of the land, they leaped into the sea and shot an infinite number of arrows at us, endeavoring might and main to prevent our landing. Their bodies were all painted over with many colors, and were decorated with birds' feathers. The natives whom we had taken with us noticed this and informed us that whenever the men are so painted and adorned with plumes they are ready for battle. They were, however, so successful in preventing our landing that we were compelled to direct our stone-hurling machines against them. When they heard the report and noticed its power (for many of them had fallen dead), they fled to the shore. We then held a consultation, and forty-two of us agreed to land after them and valiantly to engage in battle with them. This we did. We leaped to the shore fully armed; and the natives made such stout resistance that the battle raged ceaselessly for almost two hours with varying fortune. We gained a signal victory over them, but only a very few of the natives were killed, and not by us but by our cross-bowmen and gunners, which was due to the fact that they very shrewdly avoided our spears and swords. But at last we made a rush upon them with such vigor that we killed many with the points of our swords. When they saw this, and when very many had been killed and wounded, they turned in flight to the woods and forests, leaving us masters of the field. We did not wish to pursue them any further that day because we were too fatigued and preferred to make our way back to our ships. And the joy of the seven who had come with us from the mainland was so great that they could scarcely restrain themselves.

Early the next day we saw a great horde of people approaching through the island, playing on horns and other instruments which they use in war, and again painted and wearing birds' feathers. It was a wonderful sight to see. We again discussed what their plans might be, and decided upon the following course of action: to gather our forces quickly if the natives offered us any hostility; to keep constant watch in turns and in the meantime to endeavor to make them our friends, but to treat them as enemies if they rejected our friendship; and finally to capture as many of them as we could and make and keep them as our slaves forever. And so we gathered upon the shore in hollow formations, armed to the teeth. They, however, did not oppose the slightest resistance to our landing, I suppose on account of their fear of our guns. Upon disembarking, fifty-seven strong, we advanced against them in four divisions (each man under his respective captain), and engaged in a long hand-to-hand combat with them.

After a long and severe struggle, during which we inflicted great loss upon them, we put the rest to flight and pursued them as far as one of their settlements. Here we made twenty-five prisoners, set fire to the village, and returned to the ships with our captives. The losses of the enemy were very many killed and wounded; on our side, however, only one man was killed, and twenty-two were wounded, all of whom have regained their health, with the help of God.

Our arrangements for the return to our fatherland were now complete. To the seven natives who had come with us from the mainland (five of whom had been wounded in the aforesaid battle), we gave seven prisoners, three men and four women. And they, embarking in a boat which they had seized on the island, returned home filled with great joy and with great admiration for our strength. We set sail for Spain, and at last entered the harbor of Cadiz with our two hundred and twenty-two prisoners, on the 25th day of October, in the year of Our Lord 1499, where we were received with great rejoicing, and where we sold all our prisoners.

And these are what I have deemed to be the more noteworthy incidents of my first voyage.

THE SECOND VOYAGE

The following pages contain an account of my second voyage and of the noteworthy incidents which befell me in the course of that voyage.

We set sail from the harbor of Cadiz, in the year of Our Lord 1489 [sic], on a May day. As soon as we cleared the harbor, we shaped our course for the Cape Verde Island; and passing in sight of the islands of the Grand Canary group, we sailed on until we reached the island called Fire Island. Here we took on supplies of fuel and water, and resumed our voyage with a southwest wind. After nineteen days we reached a new land, which we took to be the mainland. It was situated opposite to that land of which mention has been made in our first voyage; and it is within the Torrid Zone, south of the equinoctial line, where the pole rises five degree above the horizon beyond every climate. The land is 500 leagues to the southwest of the above-mentioned islands.

We discovered that in this country the day is of the same length as the night on the 27th of June, when the sun is on the Tropic of Cancer. Moreover, we found that the country is, in great measure, marshy and that it abounds in large rivers, which cause it to have very thick vegetation and very high and straight trees. In fact, the growth of vegetation was such that we could not at the time decide whether or not the country was inhabited. We stopped our ships and anchored them, and then lowered some of our small boats in which we made for the land. We hunted long for a landing, going here and there and back and forth, but, as has already been said, found the land everywhere so covered with water that there was not a single spot that was not submerged. We saw, however, along the banks of those rivers many indications that the land was not only inhabited, but indeed very thickly populated. We could not disembark to examine such signs of life more closely, and therefore agreed to return to our ships, which we did. We weighed anchor and sailed along the coast with the wind blowing east and southeast, trying time and again, in a course of more than forty leagues, to penetrate into the island itself. But all to no purpose. For we found in that part of the ocean so strong a current flowing from southeast to northwest that the sea was quite unfit for navigation. When we discovered this difficulty, we held a council and determined to turn back and head our ships to the northwest. So we continued to sail along shore and finally reached a body of water having an outer harbor and a most beautiful island at the entrance.

We sailed across the outer harbor that we might enter the inner haven. In so doing, we noticed a horde of natives on the aforesaid island, about four leagues inland from the sea. We were greatly pleased and got our boats ready to land. While we were thus engaged, we noticed a canoe coming in from the open sea with many persons on board, which made us resolve to attack them and make them our prisoners. We therefore began to sail in their direction and to surround them, lest they might escape us. The natives in their turn bent to their paddles and, as the breeze continued to blow but moderately, we saw them raise their oars straight on high, as if to say that they would remain firm and offer us resistance. I suppose that they did this in order to rouse admiration in us. But when they became aware that we were approaching nearer and nearer, they dipped their paddles into the water and made for the land. Among our ships there was a very swift boat of about forty-five tons, which was so headed that she soon got to windward of the natives. When the moment for attacking them had come, they got ready themselves and their gear and rowed off. Since our ship now went beyond the canoe of the natives, these attempted to effect their escape. Having lowered some boats and filled them with brave men, thinking that we would catch them, we soon bore down on them, but though we pursued them for two hours, had not our caravel which had passed them turned back on them they would have entirely escaped us. When they saw that they were hemmed in on all sides by our small boats and by the ship, all of them (about twenty in number) leaped into the water, albeit they were still about two leagues out at sea. We pursued them with our boats for that entire day, and yet we managed to capture only two of them, the rest reaching land in safety.

In the canoe which they had abandoned, there were four youths, who did not belong to the same tribe, but had been captured in another land. These youths had recently had their virile parts removed, a fact which caused us no little astonishment. When we had taken them on board our ships, they gave us to understand by signs that they had been carried off to be devoured, adding that this wild, cruel, and cannibal tribe were called "Cambali."

We then took the canoe in tow, and advanced with our ships to within half a league of the shore, where we halted and dropped our anchors. When we saw a very great throng of people roaming on the shore, we hastened to reach land in our small boats, taking with us the two men we had found in the canoe that we had attacked. The moment we set foot on dry land, they all fled in great fright to the groves near by and hid in their recesses. We then gave one of the captives permission to leave us, loading him with very many gifts for the natives with whom we desired to be friends, among which were little bells and plates of metal and numerous mirrors. We instructed him, furthermore, to tell the natives who had fled not to entertain any fear on our account, because we were greatly desirous of being their friends. Our messenger departed and fulfilled his mission so well that the entire tribe, about four hundred in number, came to us from our of the forest, accompanied by many women. Though unarmed, they came to where we were stationed with our small boats, and we became so friendly that we restored to them the second of the two men whom we had captured, and likewise sent instructions to our companions, in whose possession it was, to return to the natives the canoe which we had run down. This canoe was hollowed out of the trunk of a single tree, and had been fashioned with the greatest care. It was twenty-six paces long and two ells (bracchia) wide. As soon as the natives had recovered possession of their canoe and had placed it in a secure spot along the river bank, they unexpectedly fled from us and would no longer have anything to do with us. By such an uncivilized act, we knew them to be men of bad faith. Among them we saw a little gold, which they wore suspended from their ears.

We left that country, and after sailing about eighty leagues we found a safe anchorage for our ships, upon entering which we saw such numbers of natives that it was a wonderful sight. We immediately made friends with them and visited in their company many of their villages, where we were honorably and heartily welcomed. Indeed, we bought of them five hundred large pearls in return for one small bell, which we gave them for nothing. In that land they drink wine made from fruits and seeds, which is like that made from chickpeas, or like white or red beer. The better kind of wine, however, is made from the choicest fruits of the myrrh tree. We ate heartily of these fruits and of many others that were both pleasant to the taste and nourishing, for we had arrived in at the proper season. This island greatly abounds in what they use for food and utensils, and the people themselves were well mannered and more peacefully inclined than any other tribe we met.

We spend seventeen days in this harbor very pleasantly, and each day a great number of people would come to us to marvel at our appearance, the whiteness of our skins, our clothes and weapons, and at the great size of our ships. Indeed, they even told us that one of the tribes hostile to them lived further to the west, and possessed an infinite number of pearls; and that those pearls which they themselves possessed had been taken from these enemies in the course of wars which they had waged against them. They gave us further information as to how the pearls were fished and how they grew, all of which we found to be true, as your Majesty will learn later on.

We left that harbor and sailed along the coast, on which we always saw many people. Continuing on our course, we entered a harbor for the purpose of repairing one of our ships. Here again we saw many natives, whom we could neither force nor coax to communicate with us in any way. For, if we made any attempt to land, they resisted most desperately; and if they could not withstand our attack, they fled to the woods, never waiting for us to approach any nearer. Realizing their utter savageness, we departed. While we were thus sailing on, we saw an island fifteen leagues out at sea and resolved to visit it and learn whether or not it was inhabited. Upon reaching it we found it to be inhabited by a race of most animal-like simplicity, and at the same time very obliging and kind, whose rites and customs are the following:

On the Rites and Customs of This Tribe

They were animal-like in their appearance and actions, and had their mouths full of a certain green herb which they continually chewed upon as animals chew their cud, with the result that they could not speak. Moreover, each one of them had suspended from his neck two small dried gourds, one of which contained a supply of that herb which they were chewing, while the other contained a kind of white flour resembling plaster or white lime. Every now and then they would thrust into the gourd filled with flour a small stick whose end they had moistened in their mouths. By so doing they managed to gather some of the flour and put it into their mouths, powdering with this flour that herb which they were already chewing. They repeated this process at short intervals; and though we wondered greatly, we could not see any reason for their so doing, nor could we understand their secret.

This tribe came to us and treated us as familiarly as if they had frequently had dealings with us and as if they had long been friendly with us. We strolled with them along the shore, talking the while, and expressed our desire to drink some fresh water. To which they answered, by signs, that there was none in their country, offering us in its stead some herb and flour such as they were chewing. We now understood that since their country lacked water, they chewed that herb and flour to quench their thirst. And so it happened that, though we walked along that shore in their company for a day and a half, we never came across any spring water, and learned that such water as they did drink was the dew which gathered upon certain leaves having the shape of a donkey's ears. During the night these leaves were filled with dew, of which the people then drank, and it is very good. But in many places these leaves are not found.

This tribe is entirely unacquainted with the solid products of the earth, and live chiefly on the fish which they catch in the sea. Indeed there are many expert fishermen among them, and their waters abound in fish, of which they offered us many turtles and many other most excellent varieties. The women of the tribe, however, do not chew the herb as the men do; in its place, each one of them carries a single gourd filled with water, of which they partake from time to time. They do not have villages composed of individual houses, nor do they have even small huts. Their only shelter is made of large leaves, which serve indeed to protect them against the heat of the sun, but are not a sufficient protection against the rains, from which it may be deduced that there is little rain in that country. When they come down to the sea to fish, each one brings with him a leaf so large that, by fixing one end of it in the ground and then turning the leaf to follow the sun, he procures underneath its shade ample relief from the great heat. In this island, finally, there are countless species of animals, all of which drink the water of the marshes.

Seeing, however, that there was nothing to be gained on that island, we left it and found another one. We landed and started to search for some fresh water to drink, believing the island to be uninhabited because we had seen no one as we approached it. But as we were walking along the shore, we came upon some very large footprints, from which we judged that, if the other members of the body were in proportion to the size of the feet, the inhabitants must be very large indeed. Continuing our walk along the sands, we discovered a road leading inland, along which nine of us decided to go to explore the island, because it did not seem to be very large nor very thickly populated. After advancing along that road about a league, we saw five houses situated in a valley and apparently inhabited. Entering them we found five women, two of them old and three young; and all of them were such large and noble stature that we were greatly astonished. As soon as they laid eyes upon us they were so overcome with surprise that they had not strength left for flight. Thereupon the old woman addressed us soothingly in their own tongue, and, gathering in one hut, offered us great quantities of food. All of them, in truth, were as tall as Francesco degli Albizi, and better knit and better proportioned than we are. When we had observed all this, we agreed to seize the young girls by force and to bring them to Castile as objects of wonder.

While we were still deliberating, behold about thirty-six men began to file through the door of the house, men much larger than the women and so magnificently built that it was a joy to see them. These men caused us such great uneasiness that we considered it safer to return to our ships than to remain in their company. For they were armed with immense bows and arrows, and with stakes and staffs the size of long poles. As soon as they had all entered, they began to talk among themselves as if plotting to take us prisoners, upon seeing which we, too, held a consultation. Some were of the opinion that we should fall upon them just where they were, within the hut itself; others disapproved of this entirely, and suggested that the attack be made out of doors and in the open; and still others declared that we should not force an engagement until we learned what the natives decided to do. During the discussion of these plans we left the hut disguising our feelings and our intentions, and began to make our way back to the ships. The natives followed at a stone's throw, always talking among themselves. I believe, however, that their fear was no less than ours; for, although they kept us in sight, they remained at a distance, not advancing a single step unless we did likewise. When, however, we had reached the ships and had boarded them in good order, the natives immediately leaped into the sea and shot very many of their arrows after us. But now we had not the slightest fear of them. Indeed, rather to frighten than to kill them, we shot two of our guns at them; and upon hearing the report they hastily fled to a hill nearby. Thus it was that we escaped from them and departed. These natives, like the others, also go about naked; and we called the island the Island of the Giants, on account of the great size of its inhabitants.

We continued our voyage further, sailing a little further off shore than before and being compelled to engage with the enemy every now and then because they did not want us to take anything out of their country. By this time thoughts of revisiting Castile began to enter our minds, particularly for this reason, that we had now been almost a year at sea and that we had very small quantities of provisions and other necessaries left. Even what still remained was all spoiled and damaged by the extreme heat which we had suffered. For, ever since our departure from the Cape Verde Islands, we had continually sailed in the Torrid Zone, and had twice crossed the equator, as we have said above.

While we were in this state of mind, it pleased the Holy Spirit to relieve us of our labors. For, as we were searching for a suitable haven wherein to repair our ships, we reached a tribe which received us with the greatest demonstrations of friendship. We learned, moreover, that they were the possessors of countless large Oriental pearls. We therefore remained among them forty-seven days, and bought 119 marcs of pearls at a price which, according to our estimation, was not greater than forty ducats, for we gave them in payment little bells, mirrors, bits of crystals, and very thin plates of electrum. Indeed, each one would give all the pearls he had for one little bell. We also learned from them how and where the pearls were fished, and they gave us several of the shells in which they grow. We bought some shells in addition, finding as many as 130 pearls in some, and in others not quite so many. Your Majesty must know that unless the pearls grow to full maturity and of their own accord fall from the shells in which they are born, they cannot be quite perfect. Otherwise, as I have myself found by experience time and again, they soon dry up and leave no trace. When, however, they have grown to full maturity, they drop from the fleshy part into the shell, except the part by which it hung attached to the flesh; and these are the best pearls.

At the end of the forty-seven days, then, we took leave of that tribe with which we had become such good friends, and set sail for home on account of our lack of provisions. We reached the island of Antiglia, which Christopher Columbus had discovered a few years before. Here we remained two months and two days in straightening out our affairs and repairing our ships. During this time we endured many annoyances from the Christians settled on that island, all of which I shall here pass over in silence that I may not be too prolix. We left that island on the 27th of July, and after a voyage of a month and a half we at last entered the harbor of Cadiz on the 8th of September, where we were received with great honor.

And so ended my second voyage, according to the will of God.

THE THIRD VOYAGE

I had taken up my abode in Seville, desiring to rest myself a little, to recover from the toils and hardships endured in the voyages described above, intending finally to revisit the land of pearls. But Fortune was by no means done with me. For some reason unknown to me she caused his most serene Lordship, Manuel, King of Portugal, to send me a special messenger bearing a letter which urgently begged me to go to Lisbon as soon as possible, because he had some important facts to communicate to me. I did not even consider the proposition, but immediately sent word by the same messenger that I was not feeling very well and in fact was ill at that moment; adding that, if I should regain my health and if it should still please His Royal Majesty to enlist my services, I should gladly undertake whatever he wished. Whereupon the King, who saw that he could not bring me to him just then, sent to me a second time, commissioning Giuliano Bartolomeo Giocondo, then in Lisbon, to leave no stone unturned to bring me back to the King. Upon the arrival of the said Giuliano I was moved by his entreaties to return with him to the King—a decision which was disapproved of by all those who knew me. For I was leaving Castile, where no small degree of honor had been shown me and where the King himself held me in high esteem. What was even worse was that I departed without taking leave of my host. I soon presented myself before King Manuel, who seemed to rejoice greatly at my arrival. He then repeatedly asked me to set out with three ships which had been got ready to start in search of new lands. And so, inasmuch as the entreaties of Kings are as commands, I yielded to his wishes.

The Start of the Third Voyage

We set sail in three ships from the harbor of Lisbon, on the 10th of May, 1501, directing our course toward the islands of the Grand Canary. We sailed along in sight of these islands without stopping, and continued our westward voyage along the coast of Africa. We delayed three days in these waters, catching a great number of species of fish called *Parghi*. Proceeding thence we reached that region of Ethiopia which is called Besilicca, situated in the Torrid Zone, within the first climate, and at a spot where the North Pole rises fourteen degrees above the horizon. We remained here eleven days to take on supplies of wood and of water, because it was my intention to sail southward through the Atlantic Ocean. We left that harbor of Ethiopia and sailed to the southwest for sixty-seven days, when we had reached an island 700 leagues to the southwest of the above-mentioned harbor. During these days we encountered worse weather than any human being had ever before experienced at sea. There were high winds and violent rainstorms which caused us countless hardships. The reason for such inclement weather was that our ships kept sailing along the equinoctial line, where it is winter in the month of June and the days are as long as the nights, and where our own shadows pointed always to the south.

At last it pleased God to show us new land on the 17th of August. We anchored one league and a half out at sea, and then, embarking in some small boats, we set out to see whether or not the land was inhabited. We found that it was thickly inhabited by men who were worse than animals, as Your Royal Majesty will learn forthwith. Upon landing we did not see any of the natives, although from many signs which we noticed we concluded that the country must have many inhabitants. We took possession of the coast in the name of the most serene King of Castile, and found it to be a pleasant and fruitful and lovely land. It is five degrees south of the Equator. The same day we returned to our ships; and since we were suffering from the lack of fuel and water, we agreed to land again the following day and provide ourselves with what was necessary. Upon landing we saw on the topmost ridge of a hill many people who did venture to descend. They were all naked and similar in both appearance and color to those we had met in the former voyages. Though we did our best to make them come down to us and speak with us, we could not inspire them with sufficient confidence. Seeing their obstinacy and waywardness, we returned to our ships at night, leaving on the shore (as they looked on) several small bells and mirrors and other such trifles.

When they saw that we were far out at sea, they came down from the mountain to take the things we had left them, and showed great wonder thereat. On that day we took on a supply of water only. Early in the morning of the next day, as we looked out from our ships, we saw a larger number of natives than before, building here and there along the shore fires which made a great deal of smoke. Supposing that they were thus inviting us, we rowed to the land. We now saw that a great horde of natives had collected, who, however, kept far away from us, making many signs that we should go with them into the interior. Wherefore two of our Christians declared themselves ready to risk their lives in this undertaking and to visit the natives in order to see for themselves what kind of people they were and whether they possessed any riches or aromatic spices. They begged the commander of the fleet so earnestly that he gave his consent to their departure. The two then prepared themselves for the expedition, taking along many trifles, for barter with the natives, and left us, with the understanding that they should make sure to return after five days at the most, as we should wait for them no longer.

They accordingly began their journey inland, and we returned to our ships, where we waited for eight whole days. On almost each of these days a new crowd would come to the shore, but never did they show a desire to enter into conversation with us. On the seventh day, while we again were making our way to the shore, we discovered that the natives had brought all their wives with them. As soon as we landed they sent many of their women to talk with us. But even the women did not trust us sufficiently. While we were waiting for them to approach, we decided to send to them one of our young men who was very strong and agile; and then, that the women might be the less fearful, the rest of us embarked in our small boats. The young man advanced and mingled among the women; they all stood around him, and touched and stroked him, wondering greatly at him. At this point a woman came down from the hill carrying a big club. When she reached the place where the young man was standing, she struck him such a heavy blow from behind that he immediately fell to the ground dead. The rest of the women at once seized him and dragged him by the feet up the mountain, whereupon the men who were on the mountain ran down to the shore armed with bows and arrows and began to shoot at us. Our men, unable to escape quickly because the boats scraped the bottom as they rowed, were seized with such terror that no one had any thought at the moment of taking up his arms. The natives had thus an opportunity of shooting very many arrows at us. Then we shot four of our guns at them; and although no one was hit, still, the moment they heard the thunderous report, they all fled back to the mountain. There the women, who had killed the youth before our eyes, were now cutting him to pieces, showing us the pieces, roasting them at a large fire which they had made, and eating them. The men, too, made us similar signs, from which we gathered that they had killed our two other Christians in the same manner and had likewise eaten them. And in this respect at least we felt sure that they were speaking the truth.

We were thoroughly maddened by this taunting and by seeing

with our own eyes the inhuman way in which they had treated our dead. More than forty of us, therefore, determined to rush to the land and avenge such an inhuman deed and such bestial cruelty. But the commander of our ship would not give his consent; and so, being compelled to endure passively so serious and great an insult, we departed with heavy hearts and with a feeling of great shame, due to the refusal of our captain.

Leaving that land we began to sail between the East and South because the coast line ran in that direction. We made many turns and landings, in the course of which we did not see any tribe which could have any intercourse with us or approach us. We sailed at last so far that we discovered a new land stretching out toward the southwest. Here we rounded a cape (to which we gave the name St. Vincent) and continued our voyage in a southwesterly direction. This Cape St. Vincent is 150 leagues to the southeast of the country where our Christians perished, and eight degrees south of the Equator. As we were sailing along in this manner, one day we noticed on the shore a great number of natives gazing in wonder at us and at the great size of our ships. We anchored in a safe place and then, embarking in our small boats, we reached land. We found the people much kinder than the others; for our toilsome efforts to make them our friends were at last crowned with success. We remained five days among them trading and otherwise dealing with them, and discovered large hollow reed-stalks, most of them still green, and several of them dry on the tops of the trees. We decided to take along with us two of this tribe that they might teach us their tongue; and, indeed, three of them volunteered to return to Portugal with us.

But, since it wearies me to describe all things in detail, may it suffice your Majesty to know that we left that harbor, sailing in a southwesterly direction, keeping always within sight of land, entering many harbors, making frequent landings, and communicating with many tribes. In fact, we sailed so far to the south that we went beyond the Tropic of Capricorn. When we had gone so far south that the South Pole rose thirty-two degrees above the horizon, we lost sight of the Lesser Bear, and the Great Bear itself appeared so low as to be scarcely visible above the horizon. We were then compelled to guide ourselves by the stars of the South Pole, which are far more numerous and much larger and more brilliant than the stars of our Pole. I therefore made a drawing of very many of them, especially of those of the first magnitude, together with the declinations of their orbits around the South Pole, adding also the diameters and semi-diameters of the stars themselves—all of which can be readily seen in my "Four Voyages." In the course of the voyage from Cape St. Augustine, we sailed 700 leagues—100 toward the west and 600 toward the southwest. Should any one desire to describe all that we saw in the course of that voyage, paper would not suffice him. We did not, however, discover anything of great importance with the exception of an infinite number of cassia trees and of very many others which put forth a peculiar kind of leaf. We saw, in addition, very many other wonderful things which it would be tedious to enumerate.

We had now been on our voyage for almost ten months; and seeing that we discovered no precious metals, we decided to depart thence and to roam over another portion of the sea. As soon as we had come to this conclusion, the word went to each one of our ships that whatever I should think necessary to command in conducting this voyage should be fulfilled to the letter. I therefore immediately gave a general order that all should provide themselves with fuel and water for six months, for the different captains had informed me that their ships could remain at sea only that much longer.

As soon as my orders had been obeyed, we left that coast and began our voyage to the south on the 13th of February, in other words, when the sun was approaching the equinoctial line and returning to this Northern Hemisphere of ours. We sailed so far that the South Pole rose fifty-two degrees above the horizon, and we could no longer see the stars of the Great or the Lesser Bear. For we were then (the 3rd of April) 500 leagues distant from that harbor from which we had begun our southward voyage. On this day so violent a storm arose that we were forced to gather in every stitch of canvas and to run on with bare masts, the southwest wind blowing fiercely and the sea rolling in great billows, in the midst of a furious tempest. The gale was so terrible that all were alarmed in no slight degree. The nights, too, were very long. For on the 7th of April, when the sun was near the end of Aries, we found that the night was fifteen hours long. Indeed, as your Majesty is very well aware, it was the beginning of winter in that latitude. In the midst of this tempest, however, on the 2nd of April, we sighted land, and sailed along shore for nearly twenty leagues. But we found it entirely uninhabited and wild, a land which had neither harbors nor inhabitants. I suppose it was for the reason that it was so cold there that no one could endure such a rigid climate. Furthermore, we found ourselves in such great danger and in the midst of so violent a storm that the different ships could scarcely sight one another. Wherefore the commander of the fleet and I decided that we should signal to all our shipmates to leave that coast, sail out to sea, and make for Portugal.

This plan proved to be a good and necessary one; for, had we remained there one single night longer, we should all have been lost. The day after we left, so great a storm arose that we feared we should be entirely submerged. For this reason we then made many vows to go on pilgrimages and performed other ceremonies, as is customary with sailors. The storm raged round us for five days, during which we could never raise our sails. During the same time we went 250 leagues out to sea, always getting nearer and nearer the equinoctial line, where both sea and sky became more moderate. And here it pleased God on high to deliver us from the above-mentioned dangers. Our course was shaped to the north and northeast, because we desired to make the coast of Ethiopia, from which we were then distant 1,300 leagues, sailing through the Atlantic Ocean. By the grace of God we reached that country on the 10th of May. We rested there for fifteen days upon a stretch of coast facing the south and called Sierra Leone. Then we took our course toward the Azores, which are 750 leagues from Sierra Leone. We reached them about the end of July and again rested for fifteen days. We then set sail for Lisbon, from which we were 300 leagues to the west. And at last, in the year 1502, we again entered the port of Lisbon, in good health as God willed, with only two ships. The third ship we had burned at Sierra Leone, because she was no longer seaworthy.

In this third voyage, we remained at sea for nearly sixteen months, during eleven of which we sailed without being able to see the North Star nor the stars of the Great and the Lesser Bear. At that time we steered by the star of the South Pole.

What I have related above I have deemed the most noteworthy events of my third voyage.

THE FOURTH VOYAGE

I must still relate what I saw in my third [sic] voyage. But, in truth, since I have already been tired out by the length of the preceding narratives, and since this voyage did not at all end as I had hoped, on account of an accident that befell us in the Atlantic Ocean, I may be permitted (I trust), to be somewhat brief.

We left Lisbon in six ships with the intention of exploring an island situated toward the horizon and known as Melcha. This island is famous for its wealth, because it is a stopping place for all ships coming from the Gangetic and Indian Seas, precisely as Cadiz is the port for all vessels going from east to west, or in the opposite direction, as is the case with those ships which sail hence for Calicut. This island of Melcha is further to the west than Calicut and more to the south, which we knew from the following fact: that it is situated within sight of the thirty-third degree of the Antarctic Pole.

And so, on the 10th of May, 1503, we set sail from Lisbon (as I have said above), and made for the Cape Verde Islands, where we took on some needed provisions and many other necessary stores. We remained there twelve days, and then set sail with a south wind, because the commander of the fleet, who was haughty and headstrong, issued orders that we should make for Sierra Leone, on the southern coast of Ethiopia. There was no necessity for this, and all of us were unanimously opposed to such a course; but he insisted upon it merely to impress upon us that he had been placed in command of us and the six ships. We made good speed, and just as we were at last coming within sight of our destination, so great and violent a tempest arose, and so heavy a gale began to rage, and Fortune became so unkind, that for four days we could not land in spite of the fact that we could see the coast during the whole of that time. Finally we were obliged to continue in what should have been our course from the beginning.

We therefore resumed our voyage with the Suduesius wind blowing (a wind which points between the south and the south-west), and sailed through those difficult seas for 300 leagues. In consequence we went across the Equator by almost three degrees, where land was seen by us twelve leagues off. We were greatly astonished at the sight. It was an island situated in the middle of the sea, very high and remarkable in appearance. It was no larger than two leagues in length by one in width. No man had ever been or lived on that island, and yet it was to us a most unfortunate island. Upon it the commander of our fleet lost his ship, all owing to his own obstinate mind and will. His ship struck upon a rock, sprung leaks, and sank during the night of St. Lawrence, the 10th of August. With the exception of the crew nothing was saved. The ship was of 300 tons, and the strength of our whole fleet lay in her.

While we were all exerting ourselves to see if we could not, perhaps, float her again, the above-mentioned commander ordered me (among other things) to go in a rowboat to the island in search of a good harbor where we might all draw up our ships in safety. That same commander, however, did not wish me to go with my own ship, because it was manned by nine sailors and was then busily engaged in assisting the endangered ship. He insisted that I go and find such a harbor, where he would restore my ship to me in person. Upon receiving these orders, I went to the island as he desired, taking with me about half the number of my sailors. The island was four leagues away, and hastening thither I discovered a very fine harbor where we might safely anchor our entire fleet. I had now discovered the harbor, and there I spent eight days waiting for the said commander and the rest of our company. I was greatly disturbed when they did not appear, and those who were with me became so alarmed that they could not be appeased in any way.

While we were in this state of anxiety, on the eighth day we saw a ship coming in over the sea. We at once set out to meet them in order that they might see us, feeling confident and at the same time hoping that they would take us with them to some better harbor. When we had gotten near and had exchanged greetings, those on board informed us that the commander's ship had been lost at sea, the crew alone being saved. Your Majesty can readily imagine the great anxiety which seized me at this report, when I realized that I was 1,000 leagues distant from Lisbon (to which I must needs return) in remote and far-off waters. Nevertheless, we resigned ourselves to the fate that had come upon us and determined to go on. First of all we returned to the island, where we gathered supplies of wood and water for the ship. The island, indeed, was quite uninhabited and most inhospitable; but it had a great deal of spring water, countless tree, and numberless land and sea birds, which were so tame that they permitted us to take them in our hands. We, therefore, took so many of them that we entirely filled one of the rowboats. The only other animals we discovered on that island were very large mice, lizards with forked tails and several serpents.

When we had got our provisions on board, we set sail toward the south and southwest; for we had received orders from the King, that, unless some great danger made it impossible, we should follow in the path of our former voyage. Setting out, therefore, in this direction, we at last found a harbor which we called the Bay of All Saints. Indeed, God had granted us such favorable weather that in less than seventeen days we reached this port, which is 300 leagues distant from the above-mentioned island. In the harbor we found neither the commander-in-chief nor any one else of our company, though we waited for them for two months and four days. At the end of this period, seeing that no one arrived there, my companions and I decided to sail further along the coast. After sailing for 260 leagues, we entered a harbor where we determined to build an outpost. Having done so, we left behind in this fort the twenty-four Christians who had been the crew of the luckless ship of our commander-in-chief. We remained in that harbor five months, occupied in constructing the said fort and in loading our ships with brazilwood. We tarried this long because our sailors were few in number and because, owing to the lack of many necessary parts, our ships could not proceed further. But when all was done, we agreed to return to Portugal, to do which would require a wind between north and northeast.

We left in the fort the twenty-four Christians, giving them twelve guns and many more arms, and supplying them with provisions to last them six months. During our stay we had made friends with the tribes of that country, of which we have here made very little mention, notwithstanding that we saw great numbers of them and had frequent dealings with them. Indeed, we went about forty leagues into the interior in company with thirty of them. I saw on this expedition very many things which I now pass over in silence, reserving them for my book entitled "The Four Voyages." That country is eight degrees south of the equator and thirty-five degrees west of the meridian of Lisbon, according to our instruments.

We set sail hence with the Nornordensius wind (which is between the north and the northeast) shaping our course for the city of Lisbon. At last, praise be to God, after many hardships and many dangers we entered this harbor of Lisbon in less than seventy-seven days, on the 28th of June, 1504. Here we were received with great honor and with far greater festivities than one would think possible. The reason was that the entire city thought that we had been lost at sea, as was the case with all the rest of our fleet, who had perished owing to the foolish haughtiness of our commander-in-chief. Behold the manner in which God, the just Judge of all, rewards pride!

I am now living at Lisbon, not knowing what next your most serene Majesty will plan for me to do. As for myself, I greatly desire from now on to rest from my many hardships, in the meantime earnestly commending to your Majesty the bearer of the present letter.

Amerigo Vespucci in Lisbon

ILLUSTRATIONS

Frontispiece. Martin Waldseemüller.

Figure 1. Probable site of the original Saint Dié press.

Figure 2A and B. Woodcuts from *De insulis in mari Indico nuper inventis*, Columbus's printed Letter Basel, 1493.

Figure 3. Colophon for *Cosmographiae Introductio*.

Figure 4A. Title page from *Cosmographiae Introductio*.

Figure 4B. Translation of the title page from *Cosmographiae Introductio*.

Figure 5. *Cosmographiae Introductio* page with "America(o)" in margin and text.

Figure 6A. Amerigo Vespucci (1454–1512) portrait.

Figure 6B. Statue of Amerigo Vespucci in Uffizi Gallery, Florence, Italy.

Figure 7. Vespucci family; Fresco "La Pietà e Misericordia" by Domenico Ghirlandaio in the church of San Salvadore of Ognissanti, Florence.

Figure 8. Christopher Columbus in *Americae Retectio*, Johannes Stradanus and Adrianus Collaert, Antwerp, ca. 1595.

Figure 9. Amerigo Vespucci in *Americae Retectio*, Johannes Stradanus and Adrianus Collaert, Antwerp, ca. 1595.

Figure 10. Johann Schöner (1477–1547).

Figure 11. Wolfegg Castle, modern photograph.

Figure 12. Wolfegg Castle, 1628.

Figure 13. Father Joseph Fischer, SJ.

Figure 14. Claudii Phtolomei and Ptolemaic World, sheet 2 (top row/ second from left), Waldseemüller world map of 1507.

Figure 15. Amerigo Vespucci and Western Hemisphere, sheet 3 (top row/third from left), Waldseemüller world map of 1507.

Figure 16. Stabius-Heinfogel star chart engraved by Albrecht Dürer. "Imagines coeli Meridionales," 1515.

Figure 17. Prince Waldburg-Wolfegg-Waldsee and Princess Viviana at the occasion of the display of the Waldseemüller world map of 1507 during the opening of the "Rivers, Edens, Empires: Lewis & Clark and the Revealing of America" exhibition, July 23, 2003.

ENDNOTES

CHAPTER 1

1. On the first naming of Saint-Dié-des-Vosges, see Albert Ronsin, *La Fortune d'un nom: America, Le baptême du Noveau Monde à Saint-Dié-des-Vosges* (Grenoble: Jerome Millon, 1991), p. 19. On the geographical features of the region, see John Boyd Thacher, *The Continent of America: Its Discovery and Its Baptism* (New York: William Evarts Benjamin, 1896), p. 115.

2. On the St. Dié Society, the Chicago Exposition, and Jules Ferry, see Ronsin, *La Fortune d'un nom*, pp. 83–84.

3. On the discovery of Roman walls and coins, see Ronsin, *Saint-Dié, Vosges* (Colmar-Ingersheim: S.A.E.P, 1972), p. 11.

4. On Deodatus, see Ronsin, *Les Vosgiens Célèbres* (France: Gérard Louis Vagney, 1990), pp. 110–11.

5. On Charlemagne, see Ronsin, *Saint-Dié*, p. 14.

6. On the three different groups in Saint Dié, see ibid., p. 15.

7. On the new library, see Thacher, *The Continent of America*, p. 116.

8. On Duke René and his encouragement of music, see ibid.

9. On Gaultier Lud, see Ronsin, *Les Vosgiens Célèbres*.

10. On Gaultier Lud and the *Speculum Orbis*, see Ronsin, *La Fortune d'un nom*, pp. 27–28.

11. On the *Speculi Orbis succintiss. sed neque poenitenda neqz inelegans Declaratio et Canon*, see Henry Newton Stevens, *The First Delineation of the*

New World and the First Use of the Name America on a Printed Map (London: Henry Stevens, Son & Stiles, 1928), p. 31.

12. On the three declarations contained in the book, see ibid., p. 33.

13. On the *Heures de la Vierge*, see Ronsin, *La Fortune d'un nom*, pp. 26–27.

14. Ibid., p. 27.

15. On the members of the Gymnasium Vosagense, see Thacher, *The Continent of America*, p. 116.

16. On Jean Basin de Sendacour, see Ronsin, *Les Vosgiens Célèbres*, pp. 34–35.

17. On Matthias Ringmann's education, see Ronsin, *La Fortune d'un nom*, p. 25. On Ringmann's contributions to the new edition of *Margarita Poetica*, see Thacher, *The Continent of America*, pp. 117–18.

18. On Ringmann's activities in Strasbourg, see Ronsin, *La Fortune d'un nom*, p. 15.

19. Thacher, *The Continent of America*, pp. 126–27.

20. On the *Speculis Orbis . . . Declaratio*, see Stevens, *The First Delineation*, p. 34.

21. On the notion that Ringmann was the editor of the *Cosmographiae Introductio* rather than Martin Waldseemüller, see Ronsin, *La Fortune d'un nom*, p. 27. For Ringmann's location at the beginning of March 1507, see ibid., p. 40.

22. For Ringmann's dedication to Emperor Maximilian, see Thacher, *The Continent of America*, p. 131.

23. Quote taken from ibid., p. 133.

24. Ibid., p. 156.

25. On the details of Waldseemüller's birth, see *The Catholic Encyclopedia*, online at http://www.newadvent.org/cathen/155312.htm. For a detailed description of his life, see Silvio A. Bedini, ed., *The Christopher Columbus Encyclopedia* (New York: Simon & Schuster, 1992), pp. 729–31. On Schött's activities at the Gymnasium Vosagense, see Elizabeth Harris, "The Waldseemüller World Map: A Typograph Appraisal," *Imago Mundi* 37 (1985): 49. On the matriculations of Waldseemüller and Schött, see R. A. Skelton, *Claudius Ptolemaeus Geographia Strassburg 1513 Theatrum Orbis Terrarum Ltd.*, facsimile ed. (Amsterdam: 1966), p. vii.

26. On the details of Waldseemüller's life, see Bedini, *The Christopher Columbus Encyclopedia*, p. 729.b.

27. Quote taken from Washington Irving, *The Life and Voyages of Christopher Columbus* (New York: Thomas Y. Crowell & Co., 1848), p. 646.

28. Quote taken from Thacher, *The Continent of America*, pp. 138–39.

29. On the publications at Saint Dié on and after April 25, 1507, see ibid., p. 129.

30. On the printing of the world map and the gore map, see Harris, "The Waldseemüller World Map," p. 33.

31. On the preservation of the "Carta Itineraria Europae," see Hans Wolff, ed., *America: Early Maps of the New World* (Munich: Prestel, 1992), p. 117.

32. On Waldseemüller's commitments, see Bedini, *The Christopher Columbus Encyclopedia*, p. 731.

33. Wolff, *America: Early Maps of the New World*, pp. 118–19.

34. For conflicting dates on Waldseemüller's death, see Bedini, *The Christopher Columbus Encyclopedia*, p. 731; Ronsin, *Les Vosgiens Célèbres*.

35. For the published printings at Saint Dié, see Thacher, *The Continent of America*, pp. 120–21.

36. On the transfer of the press to Strasbourg, see Thacher, *The Continent of America*, p. 121.

CHAPTER 2

1. On Columbus's undertaking, see Samuel Eliot Morison, *The European Discovery of America: The Southern Voyages, A.D. 1492–1616* (New York: Oxford University Press, 1974), p. 26.

2. On the difference between Columbus's estimate and the actual nautical distance between Europe and Japan, see ibid., p. 30. On the letter written to Columbus, see ibid., p. 27.

3. On Columbus's landing in Puerto Rico, see ibid., p. 112.

4. On Columbus's voyage, see ibid., p. 73.

5. For the attempted native uprising, see ibid., pp. 123–40.

6. For the natives' hints of land to Columbus, see Samuel Eliot Morison, *Admiral of the Ocean Sea* (Boston: Little, Brown, 1942), p. 508.

7. For the description of Soncino's description of Cabot, see Samuel Eliot Morison, *The European Discovery of America: The Northern Voyages, A.D. 600–1600* (New York: Oxford University Press, 1971), p. 158.

8. John Day's letter on Cabot quoted from W. P. Cumming, R. A. Skelton, and D. B. Quinn, *The Discovery of North America* (New York: American Heritage Press, 1972), p. 80.

9. Reference to landfall as found on an inscription on a map, Cumming et al., *The Discovery of North America*, p. 52.

10. For Cabot's sighting of an island, see Morison, *The Northern Voyages*, pp. 172–77; for Raimondo Soncino's letters, see Cumming et al., *The Discovery of North America*, p. 52.

11. For Soncino's letter, see Cumming et al., *The Discovery of North America*, p. 52; for the possible locations of Cabot's landing, see Cumming et al., *The Discovery of North America*, p. 52.

12. On the depiction of Newfoundland, see ibid., p. 52.

13. On Sebastian Cabot being an unreliable source, see Henry Harrisse, *The Discovery of North America* (Amsterdam: N. Israel, 1961), p. 195; on the questionable accuracy of the map, see ibid., p. 74.

14. For the explorers landing on the south shore of the Paria Peninsula, see Morison, *Admiral*, p. 541. On the naming of "la ysla de Trinidad," see Frederick J. Pohl, *Amerigo Vespucci: Pilot Major* (New York: Columbia University Press, 1944), p. 44 and Morison, *Admiral*, p. 528.

15. For the naming of Margarita Island, see Morison, *Admiral*, p. 555; on the naming of Asunción, see Morison, *The Southern Voyages*, p. 153; on Columbus's recordings in his journal, see Morison, *Admiral*, pp. 154–55.

16. On Vespucci's life, see Morison, *The Southern Voyages*, p. 185.

17. For the ships' journey and landing, see ibid., pp. 187–89.

18. On Bastidas's journey from Seville, see ibid., p. 199.

19. On the death of Hojeda, see ibid., p. 193.

20. Morison, *The Northern Voyages*, p. 215.

21. Morison, *Admiral*, p. 589.

22. On the naming of the coast of Honduras, see ibid., p. 595; for Columbus setting foot on land at Cape Honduras, see ibid., p. 596.

23. Ibid., p. 658.

24. For Cuneo's letter and Columbus's memorial addressed to the Spanish monarchy, see Harrisse, *The Discovery of North America*, pp. 403–404; on Blanes's map, see ibid., p. 505.

25. For the quotes from Soncino, Ayala, and Puebla, see ibid., pp. 407–409.

26. For the quotes from Hojeda and Ibarra, see ibid., pp. 408–409.

27. On Las Casas's quote and Harrisse's opinion on the authorship of the map, see ibid., p. 409.

28. On Irving's publication, see George Emra Nunn, *The Mappemonde of Juan de la Cosa; A Critical Investigation of Its Date* (Jenkintown, PA: George H. Beans Library, 1934), p. 2; on Humboldt's role in the map, see ibid., p. 2.

29. On Henry Stevens's offer bid for the map, see Henry Stevens, *Recollections of James Lenox and the Formation of His Library*, rev. and elucidated by Victor Hugo Palsits (New York: New York Public Library, 1951), pp. 50–51; for Sagra's willingness to pay more for the map, see Nunn, *The Mappemonde*, p. 1.

30. For the notion that Juan de la Cosa sailed on Columbus's first and second voyages, see Harrisse, *The Discovery of North America* and Nunn, *The Mappemonde*, p. 1; for Morison's contrasting viewpoint, see Morison, *Admiral*, pp. 144–45.

31. On the details of the map, see Nunn, *The Mappemonde*, p. 49.

32. For Nunn's argument about the map, see ibid., pp. 51–52.

33. Emerson Fite and Archibald Freeman, *A Book of Old Maps Delineating American History* (New York: Dover, 1969), p. 12.

34. Ibid.

35. For the declaration statement, see Harrisse, *The Discovery of North America*, p. 416.

36. For quotation, see Pohl, *Amerigo Vespucci*, p. 192.

37. For Pinzón's assertion and d'Anghiera's argument about Bishop Juan de Fonseca, see Harrisse, *The Discovery of North America*, pp. 416–17.

38. For Pinzón's reference to other maps and narratives that indicated Columbus had access to the marine charts of others, see ibid., pp. 419–20.

39. For Cantino's letter, see ibid., p. 77; on the cost of making the map, see ibid., p. 422.

40. For Signor Boni's actions to save the map, see ibid., pp. 423–24.

41. For translation of quote, see ibid., p. 68.

42. On Harisse's opinion of Florida, see ibid., p. 131; for Stevenson's explanation, see Edward Luther Stevenson, *Marine World Chart of Nicolo de Canerio Januensis* (New York: Hispanic Society of America, 1908), p. 32.

43. For the 1890 announcement and the map, see Harrisse, *The Discovery of North America*, pp. 428–29; on the Genoese man, see Kenneth Nebenzahl, *Rand McNally Atlas of Columbus and the Great Discoveries* (Chicago: Rand McNally & Co., 1990), p. 40.

44. For quote, see Harrisse, *The Discovery of North America*, p. 429.

45. For the belief in the antecedent of the 1507 map, see ibid., p. 430; for replication of the Florida map, see Nebenzahl, *Rand McNally Atlas*.

46. For quote, see Harrisse, *The Discovery of North America*, p. 429.

47. For Fischer and von Wieser's analysis, see J. Fischer and F. von Wieser, *The World Maps of Waldseemüller (Ilacomilus) 1507 & 1516* (Innsbruck: Verlag der Wagnerschen Universitäts-Buchhandlung, 1903), pp. 28–29; for Stevenson's agreement with Fischer and von Wieser, see Stevenson, *Marine World*, p. 82.

48. For Bartholomew Columbus's map and Martyr's report, see Harrisse, *The Discovery of North America*, pp. 436–37; for Professor von Wieser's work, see Franz R. von Wieser, *Die Carte des Bartolomeo Columbo über die vierte Reise des Admirals* (Innsbruck, 1893), p. 13.

49. For Columbus's letter to the king, see Fite and Freeman, *A Book of Old Maps*, p. 15.

50. For the inscription, see ibid., p. 19.

51. For the map inscription, see ibid.

52. Stevens, *Recollections of James Lenox*, p. 121.

53. For Dickinson's theory, see *Washington Post*, October, 7, 2002, p. A7.

54. Harrisse, *The Discovery of North America*, p. 438.

55. Gavin Menzies, *1421: The Year China Discovered America* (New York: HarperCollins, 2003).

56. Ibid., p. 37.

57. Ibid., p. 495.

58. Ibid., pp. 107, 359, 505.

59. Ibid., pp. 242, 316, 367, 501.

60. Ibid., p. 505.

61. Ibid., p. 416.

62. Quote taken from Hans Wolff, *America: Early Maps of the New World* (Munich: Prestel, 1992), p. 114.

63. Christine R. Johnson, "Renaissance German Cartographers and the Naming of America," *Past & Present* 191, no. 1 (Oxford: 2006): 3–43.

64. John Hessler, "Warping Waldseemüller: A Phenomenological and Computational Study of the 1507 World Map" *Cartographica* 4 (2006) 101–13.

65. Ibid.

CHAPTER 3

1. On the description of the *Cosmographiae Introductio*, see John Boyd Thacher, *The Continent of America: Its Discovery and Its Baptism* (New York: William Evarts Benjamin, 1896), p. 129.

2. Quote taken from Washington Irving, *The Life and Voyages of Christopher Columbus* (New York: Thomas Y. Crowell & Co., 1848), p. 644.

3. On the brief history of the book, see L. Kellner, *Alexander von Humboldt* (London: Oxford University Press, 1963), p. 180; for the final resting place of the book, see Thacher, *The Continent of America*, p. 130.

4. Thacher, *The Continent of America*, p. 130.

5. For Waldseemüller's dedication to the emperor, see ibid., p. 139.

6. On the differences between Thacher and other scholars, see Thacher, *The Continent of America*, pp. 143–44; Frederick J. Pohl, *Amerigo Vespucci: Pilot Major* (New York: Columbia University Press, 1944), p. 157.

7. Albert Ronsin, *La Fortune d'un nom: America* (Grenoble: Jerome Millon, 1991), pp. 44–46.

8. Thacher, *The Continent of America*, p. 146.

9. Christine R. Johnson, "Renaissance German Cartographers and the Naming of America," *Past & Present* 191, no. 1 (Oxford, 2006): 3–43.

10. For the unlikely letter sent to the duke, see J. Fischer and F. von Wieser, *The World Maps of Waldseemüller (Ilacomilus) 1507 & 1516* (Innsbruck: Verlag der Wagnerschen Universitäts-Buchhandlung, 1903), p. 11.

11. On the description of the geography that resembled Cape Gracias a Dios in Honduras, see Pohl, *Amerigo Vespucci*, p. 153.

CHAPTER 4

1. For a more complete discussion of the title "The Greatest Misnomer on Planet Earth," see Seymour I. Schwartz, "The Greatest Misnomer on Planet Earth," *Proceedings of the American Philosophical Society* 146 (2002): 264–81.

2. On the naming of Asia in the Qur'an and the view of modern etymologists, see ibid., p. 264.

3. On Vespucci's birth, see Germán Arciniegas, *Why America? 500*

Years of a Name (Bogotá, Colombia: Villegas Editores, S. A., 2002), pp. 27–30.

4. For the origins of Vespucci's parents, see Luciano Formisano, ed., *Letters from a New World* (New York: Marsilio, 1992), p. xxxv; Ross King, *Michelangelo and the Pope's Ceiling* (New York: Walker & Co., 2003), p. 22.

5. On Vespucci being on tax register for Florence, see Arciniegas, *Why America*, p. 110, n. 114.

6. On Vespucci's interest in geography, see ibid., pp. 186–88.

7. For Berardi's role as liaison between Columbus and the Spanish monarchy, see ibid., p. 232.

8. For Vespucci's role on Hojeda's voyage, see Frederick J. Pohl, *Amerigo Vespucci: Pilot Major* (New York: Columbia University Press, 1944), p. 49.

9. On Vespucci's travels, see ibid., pp. 53, 116.

10. For the reference to Vespucci's voyage in a letter, see Arciniegas, *Why America*, p. 414; for other references, see ibid., p. 419 and Pohl, *Amerigo Vespucci*, pp. 186–88.

11. For the quote on the natives, see Formisano, *Letters*, pp. 50–51.

12. On the impossible sailing, see Pohl, *Amerigo Vespucci*, p. 149.

13. Stefan Zweig, *Amerigo: A Comedy of Errors* (New York: Viking Press, 1942), p. 38

14. On Vespucci's location during the fourth voyage, see Formisano, *Letters*, p. xxiii.

15. For the history of Montalboddo's text, see Formisano, *Letters*, p. xvi.

16. Arciniegas, *Why America*, p. 478.

17. For the contents of Vespucci's letter, see Formisano, *Letters*, p. 171.

18. For Vespucci's description of the Guiana stream, see ibid., p. 172.

19. Quote taken from R. A. Skelton, *Claudius Ptolemaeus Geographia Strassburg 1513 Theatrum Orbis Terrarum Ltd.*, facsimile ed. (Amsterdam: 1996), p. xx.

20. Ibid.

21. Letter to Johann Amerbach taken from Henry Newton Stevens, *The First Delineation of the New World and the First Use of the Name America on a Printed Map* (London: Henry Stevens, Son & Stiles, 1928), p. 30.

22. On the distinction between modern geography and ancient geography, see Skelton, *Claudius Ptolemaeus Geographia*, p. 5.

23. On the Schöner globe, see Henry Harrisse, *Americus Vespuccius: A*

Critical and Documentary Review, Two Recent English Books concerning That Navigator (London: 1895), pp. 484–86.

24. Quote taken from Pohl, *Amerigo Vespucci*, p. 162.

25. On Ferdinand procuring a letter, see ibid., p. 157.

26. Quotes taken from Bartolomé de Las Casas, *History of the Indies*, ed. and trans. Andrée Collard (New York: Harper & Row, 1971), p. 62.

27. Quote taken from Pohl, *Amerigo Vespucci*, p. 175.

28. Quote taken from Arciniegas, *Why America*, p. 468.

29. Zweig, *Amerigo*, p. 85.

30. On Navarrete's work and Harrisse's criticism, see Arciniegas, *Why America*, pp. 469–75.

31. Quote taken from Ralph Waldo Emerson, *English Traits: The Complete Works of Ralph Waldo Emerson* (Cambridge: Riverside Press, 1876), pp. I, 152.

32. On Force's letter, see Manning Ferguson Force, "Some Observations on the Letters of Amerigo Vespucci, Cincinnati, 1885," read before the Congrès international des américanistes, September 1879; on the impact of the reports printed in *Lettera*, see Pohl, *Amerigo Vespucci*, p. 158; Harrisse's quote taken from Harrisse, *Americus Vespuccius*, p. 741.

33. On Magnaghi's assertions, see Alberto Magnaghi, *Amerigo Vespucci. Studio critico con speciale riguardo ad una nuova valutazione delle fonti e con documenti inediti tratti dal Codice Vaglienti (Ricciardiano 1910)* (Rome: 1926), "Una supposta lettera inedita di Amerigo Vespucci sopra il suo terzo viaggio," in *Bollettino della R. Società Geografica Italiana* II, ser. VII (1937): 589–632, and "Ancora a propisito di una nuova supposta lettera di Amerigo Vespucci sopra un suo terzo viaggio," in *Bollettino della R. Società Geografica Italiana* III, ser. VII (1938): 685–703; on Vespucci's forgeries, intentions, and reasons, see Pohl, *Amerigo Vespucci*, pp. 151–52.

34. George Tyler Northrup, *Mundus Novus. Letter to Pietro di Medic*, trans. George Tyler Northrup (Princeton, NJ: 1916).

35. For Northrup's evidence, see Pohl, *Amerigo Vespucci*, p. 229; for statement in *Descobridores do Brasil*, see Arciniegas, *Why America*, p. 475; on the labeling of documents "para-Vespuccian," see Formisano, *Letters*, p. xxxv.

36. Washington Irving, *The Life and Voyages of Christopher Columbus* (New York: Thomas Crowell & Co., 1848), p. 645.

37. On the first application of the term "American," see Martin Fro-

bisher, "The Third Voyage of *Captain Frobisher*, pretended for the discovery of *Catai*, by *Meta Incognita Anno Do. 1578*," in Richard Hakluyt, *The Third and Last Volume of the Voyages, Navigations, Traffiques and Discoveries of the English Nation* (London: George Bishop and Robert Baker, 1600), p. 93.

38. For the Germanic names of America, see "America's Baptismal Certificate," German American Heritage Society, December 2001.

39. For the alternate naming of America, see Rodney Broome, *Terra Incognita: The True Story of How America Got Its Name* (Seattle, WA: Educare Press, 2001); Albert Ronsin, *La Fortune d'un nom: America* (Grenoble: Jerome Millon, 1991), p. 80.

40. On Cabot's voyage, see Broome, *Terra Incognita*, pp. 38–39.

41. On Amerike's activities in Bristol, see ibid., p. 106.

42. Quote taken from ibid., p. 122.

43. Quote taken from ibid., pp. 138–39.

44. On the refutation of the Bristol/American connection, see Ronsin, *Découverte et baptême de l'Amerique*, 2nd ed. (Paris: Editions de l'est, 1992), pp. 202–208.

CHAPTER 5

1. For Washington Irving's comment on the 1507 map, see *The Life and Voyages of Christopher Columbus* (New York: Thomas Y. Crowell & Co., 1848), p. 644.

2. For Trithemius's letter to Monapius, see Henry Harrisse, *The Discovery of North America* (Amsterdam: N. Israel, 1961), p. 445, n. 156; for Harrisse's belief on Trithemius's reference to the 1507 world map, see Harrisse, *The Discovery of North America*, pp. 444–45; J. Fischer and F. von Wieser, *The World Maps of Waldseemüller (Ilacomilus) 1507 & 1516* (Innsbruck: Verlag der Wagnerschen Universitäts-Buchhandlung, 1903), pp. 15–16.

3. On Glareanus's attesting to the existence of the 1507 map, see Hans Wolff, ed., *America: Early Maps of the New World* (Munich: Prestel, 1992), p. 121.

4. On the relationship between Glareanus's manuscript and the 1507 map, see ibid., pp. 121–22, and Fischer and von Wieser, *The World Maps of Waldseemüller*, p. 9.

5. Quote taken from Harrisse, *The Discovery of North America*, p. 279.

6. Edward Luther Stevenson, "Terrestrial and Celestial Globes: Their History and Construction," *Hispanic Society of America* 1 (1921): 70.

7. On the Lenox globe, see ibid., p. 73.

8. On the Hauslab globe, see ibid., p. 75.

9. On the 1515 globe, see ibid., p. 83; Fischer and von Wieser, *The World Maps of Waldseemüller*, p. 5.

10. Quote taken from Fischer and von Wieser, *The World Maps*, pp. 17–18.

11. All information on Wolfegg Castle is taken from *Schnell Kunst-führer*, no. 1733 (Munich and Zurich: Schnell & Steiner), 1983.

12. Irving, *The Life and Voyages of Christopher Columbus*, pp. 9–10.

13. Quote taken from J. Fischer, *The Discoveries of the Norsemen in America with Special Relation to Their Early Cartographical Representation*, trans. Basil Soulsby (London: Henry Stevens, Son & Stiles, 1903), pp. vi–vii.

14. Ibid., p. 88.

15. On Fischer's travels and discoveries, see Wilfried Haller, "Pro. Dr. h.c.P. Josef Fischer S.J. Leben und Werk," unpublished in Archives Ignatiushaus Munich, ca. 1980.

16. On Fischer's research between 1914 and 1939, see ibid.

17. On the Vínland map, see Kirsten A. Seaver, *Maps, Myths, and Men: The Story of the Vinland Map* (Stanford, CA: Stanford University Press, 2004), pp. 86–107.

18. On Seaver's postulation of Fischer's creation of the Vínland map, see ibid., chap. 9, "The Vinland Map as a Human Creation."

19. Quote taken from Seaver, *Maps*, p. 296.

20. Quote taken from ibid., p. 262.

21. For the evidence linking Fischer with the name "isolanda Ibernica," see ibid., p. 300.

22. For the suggestion that the writing on the map was made by a more skilled hand, see ibid., p. 121.

CHAPTER 6

1. Quote on 1507 and 1516 map, taken from J. Fischer and F. von Wieser, *The World Maps of Waldseemüller (Ilacomilus) 1507 & 1516* (Innsbruck: Verlag der Wagnerschen Universitäts-Buchhandlung, 1903), p. 6.

2. Ibid., p. 17.

3. Elizabeth Harris, "The Waldseemüller World Map: A Typographic Appraisal," *Imago Mundi* 37 (1985): 30.

4. Ibid., p. 31.

5. For Fischer's letters to Princess Sophie, see Kirsten A. Seaver, *Maps, Myths, and Men: The Story of the Vinland Map* (Stanford, CA: Stanford University Press, 2004), p. 311, n. 50.

6. Harris, "The Waldseemüller World Map," p. 32.

7. Ibid., p. 47.

8. The condition and examination of the map was discussed in a personal communication with Heather Wanser.

9. Fischer's comment on map taken from Henry Harrisse, *The Discovery of North America* (Amsterdam: N. Israel, 1961), p. 30.

10. On Ptolemy's methods, see R. V. Tooley, *Maps and Map-Makers* (New York: Bonanza Books, 1961), p. 5.

11. Quote on fly taken from James Hall, *Dictionary of Subjects and Symbols in Art* (New York: Harper & Row, 1974), p. 126.

12. On the map legends, see Fischer and von Wieser, *The World Maps*, p. 25.

13. On the evidence of Ptolemaic sources, see ibid., pp. 25–26.

14. For the analysis of place names, see ibid., pp. 55–56.

15. For the meaning of "Parias," see Hans Wolff, ed., *America: Early Maps of the New World* (Munich: Prestel, 1992), p. 114.

16. On the history of the first gore map, see Fischer and von Wieser, *The World Maps*, p. 14; Wolff, *America*, p. 112; the details on the final purchase of the map were discussed in a personal communication between the author and James Ford Bell.

17. Bell's collection at the University of Minnesota is discussed in *Libraries and Culture* 34 (1999): 400–401.

18. On the relationship between the book and the map, see Harrisse, *The Discovery of North America*, pp. 467–68.

19. On the Hauslab collection, see Margrit B. Krewson, *The German Collections of the Library of Congress: Chronological Development* (Washington, DC: Library of Congress, 1994).

20. The history of the 1486 copy is taken from Wolff, *America*, pp. 7–8.

21. The details of Dr. Sack's discovery were discussed in a personal communication between the author and Ralf Eiserrmann, head of the public library in Offenburg.

22. The location of the copy was discussed in a personal communication between the author and Ralf Eiserrmann; see the above endnote.

23. On the relationship between the 1507 world map and the gore maps, see Fischer and von Wieser, *The World Maps*, pp. 14–15.

24. For the description of the Wolfegg volume, see ibid., p. 5.

25. Schöner's quote taken from Edward Luther Stevenson, "Terrestrial and Celestial Globes: Their History and Construction," *Hispanic Society of America* 1 (1921): 84–85.

26. On the "Brasilie Regio" and the Southern Strait, see ibid., pp. 83–85.

27. On the purchase of the volume, see J. Fischer, *The Discoveries of the Norsemen in America with Special Relation to Their Early Cartographical Representation*, trans. Basil Soulsby (London: Henry Stevens, Son & Stiles, 1903).

28. On the collaboration on the work, see Deborah J. Warner, *The Sky Explored: Celestial Cartography, 1500–1800* (New York: A. R. Liss, 1979), p. 71.

29. Quote taken from Fischer and von Wieser, *The World Maps*, p. 29.

30. On the agreement between the "Carta Marina" and the Caveri charts, see Fischer and von Wieser, *The World Maps*, p. 32.

31. On the congruence between the "Carta Marina" and the Caveri charts, see ibid., pp. 29–30.

32. Quote taken from ibid., p. 19.

33. On the lower shield of the map, see ibid., p. 20.

CHAPTER 7

1. On Steven's life, see "Henry Stevens," in *Dictionary of American Biography*, vol. II, 1963, pp. 611–12.

2. Henry Stevens to George Winship Parker, October 26, 1901, in *John Carter Brown Library Collections*, Providence, RI.

3. Henry Stevens to George Winship Parker, November 3, 1901, in ibid.

4. Henry Stevens to Henry Vignaud, November 3, 1901, in ibid.

5. Henry Stevens to George Winship Parker, November 9, 1901, in ibid.

6. Henry Stevens to George Winship Parker, November 11, 1901, in ibid.

7. Henry Stevens to John Carter Brown Library, December 21, 1901, in ibid.

8. Henry Stevens to John Carter Brown Library, January 1902, in ibid.

9. Ibid.

10. Henry Stevens to John Carter Brown Library, February 1902, in ibid.

11. Henry Stevens to Mr. Brown, February 1902, in ibid.

12. Mrs. Augusta Brown to George Winship Parker, March 5, 1902; Henry Stevens to George Winship Parker, March 19, 1902, in ibid.

13. Quote taken from Henry Newton Stevens, *The First Delineation of the New World and the First Use of the Name America on a Printed Map* (London: Henry Stevens, Son & Stiles, 1928), p. xii.

14. Quote taken from a 1907 Henry Stevens, Son & Stiles catalog advertising Fischer and von Wieser's work.

15. Advertisement, Henry Stevens, Sons & Stiles, 1903.

16. "Rare American for Sale," advertisement, Henry Stevens, Sons & Stiles, 1907.

17. "Map Is America's Birth Certificate," UPI, August 21, 2001.

18. Charles Heinrich to Count Wolfegg, 1912, in *Wolfegg Castle Files*.

19. 1912 letters, in *Wolfegg Castle Files*.

20. Stevens, *The First Delineation*, p. ix.

21. Henry Stevens to Henry Harrisse, September 6, 1900, in Henry Harrisse Collection, Rare Books, Library of Congress.

22. "John Carter Brown," in *Dictionary of American Biography*, vol. II, pp. 136–37.

23. Sophia Brown to George Winship Parker, February 28, 1901, in *John Carter Brown Library*.

24. Franz R. von Wieser, "Die älteste Karte mit dem Namen 'America' a. d. J. 1507 und die Carta Marina a. d. J. 1516 des Martin Waldseemüller" (The Oldest Map with the Name America of the Year 1507 and the Carta

Marina of 1516 of Martin Waldseemüller), *Petermann's Mittheilungen* 47 (1901): 271–75.

25. Basil H. Soulsby, "The First Map Containing the Name America," *Geographical Journal* 19 (1902): 201–208; Henry Harrisse Collection, Rare Books, Library of Congress.

26. On printing of Contarini's map, see Stevens, *First Delineation*, pp. xiv–xv.

27. Quote taken from Edward Heawood, "A Hitherto Unknown World Map of A.D. 1506," *Geographical Journal* LXII (1923): 279–93.

28. Stevens, *First Delineation*, pp. 3–4.

29. Ibid., pp. 6–13.

30. Ibid., pp. 14–18.

31. On Waldseemüller's quote, see C. Schmidt, "Mathias Ringmann-Philesius, Humaniste alsacien et lorrain," *Mémoires de la Societé d' Archéologie lorraine* 3 (1875): 227; for Stevens quote, see Stevens, *First Delineation*, p. 33.

32. Stevens, *First Delineation*, p. 90.

33. On Contarini's map, see ibid., pp. 115–21.

34. Quote taken from ibid., pp. 118–99.

35. Louis C. Karpinski, "The First Map with the Name America," *Geographical Review* 20 (1930): 664–68.

36. Letter quoted from Stevens, *First Delineation*, pp. 29–30.

37. Karpinski, "The First Map."

38. Quote taken from ibid., p. 668.

CHAPTER 8

1. Mr. Peitz to Karpinski, 1931, in *William L. Clements Library Files*, Ann Arbor, University of Michigan.

2. Edwin Wolff, *Rosenbach: A Biography* (New York: World Publishing, 1960), p. 382.

3. Erwin Raisz, 1938 letter, in *Wolfegg Castle Files*.

4. Ralph Ehrenberg communiqué, in files from Geography and Map Division, Library of Congress.

5. Annual Report, 1904, Library of Congress.

6. Library of Congress Archives.

7. Ibid.

8. Excerpts taken from "The Lord of the Manor and His Inheritance: An Interview with Johnanes Count Waldburg," *Baden Württemberg* 4 (1997).

9. Daniel Boorstin to Count Wolfegg, March 1992, in *Wolfegg Castle Files.*

10. Prince Waldburg-Wolfegg to Mr. Tabb, July 2001, in Library of Congress Archives.

11. Dr. Jörn Günther to Winston Tabb, May 25, 1999, in Library of Congress Archives.

12. E-mail correspondence between Philip Burden and author, July 15, 2003.

13. Library of Congress Archives.

14. Prince Waldburg-Wolfegg to Dr. Billington, June 17, 1999, in Library of Congress Archives.

CHAPTER 9

1. Quote taken from James Conway, *America's Library: The Story of the Library of Congress, 1800–2000* (New Haven, CT: Yale University Press, 2000), p. 89.

2. Quote taken from John Y. Cole, *For Congress and the Nation: A Chronological History of the Library of Congress* (Washington, DC: Library of Congress, 1979), pp. 3–4.

3. For the contents of the early library, see Louis De Vorsey Jr., *Keys to the Encounter* (Washington, DC: Library of Congress, 1992), p. 173; on the arrival of the contents, see Conway, *America's Library,* p. 13; on the rules concerning the use of the library, see Cole, *For Congress,* p. 4.

4. For the 1801 catalog, see De Vorsey, *Keys,* p. 173, and for the 1812 catalog, see Cole, *For Congress,* p. 6.

5. On the purchase and contents of Jefferson's library, see Cole, *For Congress,* p. 9, and De Vorsey, *Keys,* p. 173; on the limitation of using the Library of Congress, see Cole, *For Congress,* p. 12.

6. For the comment in the *National Journal,* see Conway, *America's Library,* p. 35.

7. On the failed purchase of Obidiah Rich's collection, see ibid., p. 38.

8. For the act authorizing the creation of the Smithsonian Institution, see Cole, *For Congress*, p. 19.

9. On the 1851 fire and reconstruction of the library, see Cole, *For Congress*, pp. 22–24; on the official report of 1852, see John A. Wolter et al., "A Brief History of the Library of Congress Geography and Map Division, 1897–1978," in *The Map Librarian in the Modern World: Essays in Honor of Walter Ristow*, ed. Helen Wallis and Lothar Zögner, presented by the IFLA Section of Geography and Map Libraries (New York: K. G. Saur, 1979), p. 56.

10. On Spofford's transformation of the Library of Congress, see Cole, *For Congress*, pp. 28–29, and Conway, *America's Library*, p. 69.

11. Spofford's quote taken from De Vorsey, *Keys*, p. 184.

12. On Putnam's acquisitions, see Conway, *America's Library*, pp. 109–10.

13. For the act establishing the National Archives, see Cole, *For Congress*, p. 98.

14. On the naming of the new building, see De Vorsey, *Keys*, p. 173.

15. Margrit B. Krewson, *The German Collections of the Library of Congress: Chronological Development* (Washington, DC: Library of Congress, 1994).

16. On the acquisition of the Bismarck material, see Krewson, *German Collections*.

17. On the 1904 Weber acquisition, see Krewson, *German Collections*.

18. On the separate section given for the map collection, see Wolter et al., "A Brief History of the Library of Congress Geography and Map Division, 1897–1978," p. 47.

19. On the expansion of the Library of Congress of the period, see ibid., p. 52.

20. On the inception of the Hall of Maps, see ibid., p. 55; for the 1901 publication of *A List of Maps*, see Cole, *For Congress*, p. 71.

21. For the pre-1600 collection in the Library of Congress, see Ralph E. Ehrenberg et al., compiler, *Library of Congress Geography and Maps: An Illustrated Guide* (Washington, DC: Library of Congress, 1996), pp. 11–12.

22. For the seventeenth- and eighteenth-century collections, see ibid., p. 25.

23. On the maps mentioned in the text, see Walter W. Ristow, compiler, *A la Carte: Selected Papers on Maps and Atlases* (Washington, DC: Library of Congress, 1972), pp. 39–44.

24. On John Mitchell, see *Dictionary of American Biography*, vol. III, ed. Dumas Malone, pp. 50–51.

25. On the acquisition of Hotchkiss's maps and papers, see Ristow, *A la Carte*, pp. 183–88.

26. For the recent acquisitions, see Ehrenberg et al., *Library of Congress Geography and Maps*, p. 5.

27. On the architectural design of the Library of Congress, see Herbert Small, *The Library of Congress: Its Architecture and Decoration* (New York: Norton, 1982), p. 31.

28. For the appearance of the term "America," see Henry Harrisse, *The Discovery of North America* (Amsterdam: N. Israel, 1961), pp. 504–505.

EPILOGUE

1. Lucia Perillo, *The Oldest Map with the Name America* (New York: Random House, 1999), pp. 119–25.

2. Edward P. Jones, *The Known World* (New York: HarperCollins, 2003), pp. 174–75.

BIBLIOGRAPHY

Arciniegas, Germán. *Why America? 500 Years of a Name.* Bogotá, Columbia: Villegas Editores, S. A., 2002.

Bedini, Silvio A., ed. *The Christopher Columbus Encyclopedia.* Vol. 2. New York: Simon & Schuster, 1992.

Boorstin, Daniel. *The Discoverers.* New York: Random House, 1963.

Broome, Rodney. *Terra Incognita: The True Story of How America Got Its Name.* Seattle, WA: Educare Press, 2001.

Casas, Bartolomé de Las. *History of the Indies.* Edited and translated by Andrée Collard. New York: Harper & Row, 1971.

The Catholic Encyclopedia. Vol. XV. New York: Robert Appleton Company, 1912.

Cole, George Watson, ed. *A Catalogue of Books relating to the Discovery and Early History of North and South America. Forming a Part of the Library of E.D. Church.* New York: Peter Smith, 1951.

Cole, John Y. *For Congress and the Nation: A Chronological History of the Library of Congress.* Washington, DC: Library of Congress, 1979.

Conway, James. *America's Library: The Story of the Library of Congress 1800–2000.* New Haven, CT: Yale University Press, 2000.

Cumming, W. P., R. A. Skelton, and D. B. Quinn. *The Discovery of North America.* New York: American Heritage Press, 1972.

d'Avezac, M. *Martin Hylacomylus Waltzenmüller, ses ouvrages et sus collaborateurs voyages d'exploration et de decouvertes.* Paris, 1867.

De Vorsey, Louis, Jr. *Keys to the Encounter*. Washington, DC: Library of Congress, 1992.

Dickson, Peter W. "A Secret First." *Mercator's World* 7 (2002): 46–51.

Dictionary of American Biography. Vol. II. 1963.

Ehrenberg, Ralph E., et al., compiler. *Library of Congress Geography and Maps: An Illustrated Guide*. Washington, DC: Library of Congress, 1996.

Emerson, Ralph Waldo. *English Traits: The Complete Works of Ralph Waldo Emerson*. Cambridge: Riverside Press, 1876.

Fischer, Joseph. "Claudius Clavus, the First Cartographer of America." *Historical Records and Studies* 6 (1913).

———. "Das älteste Stadium der Weltkarte des Joh. Ruysch (1508), mit Faksimile." *Schweizerisches Gutenbermuseum, Jahrgang* 17 (1931).

———. "Die handschriftliche Überlieferung der Ptolemäuskarten, Verhandlungen des 18. Deustschen Geographentages zu Innsbruck in." *Petermann's Mittheilungen* 58 (1912).

———. *The Discoveries of the Norsemen in America with Special Relation to Their Early Cartographical Representation*. Translated by Basil Soulsby. London: Henry Stevens, Son & Stiles, 1903.

———. *Map of the World by Jodocus Hondius 1611*. New York: Edward L. Stevenson, 1907.

Fischer, J., and F. von Wieser. *The World Maps of Waldseemüller (Ilacomilus) 1507 & 1516*. Innsbruck: Verlag der Wagnerschen Universitäts-Buchhandlung, 1903.

Fite, Emerson D., and Archibald Freeman. *A Book of Old Maps Delineating American History*. New York: Dover, 1969.

Force, Manning Ferguson. "Some Observations on the Letters of Amerigo Vespucci. Cincinnati, 1885." Read before the Congrès international des américanistes, September 1879.

Formisano, Luciano, ed. *Letters from a New World*. New York: Marsilio, 1992.

Frobisher, Martin. "The Third Voyage of Captain Frobisher, pretended for the discovery of Catai, by Meta Incognita Anno Do. 1578." In Richard Hakluyt, *The Third and Last Volume of the Voyages, Navigations, Traffiques and Discoveries of the English Nation*. London: George Bishop and Robert Baker, 1600.

Fuhrmann, Otto W. *The 500th Anniversary of the Invention of Printing*. New York: Philip C. Duschness, 1937.

Gallois, L. "Le portulan de Nicolas de Canerio." *Bulletin de la Société de Géographie de Lyon* (1890).

———. "Waldseemüller chanoine de Saint-Dié." *Bulletin de la Société de Geographie de l'Est* (1890): 221–29.

German American Heritage Society of Washington, DC. "America's Baptismal Document." December 2001.

Hall, James. *Dictionary of Subjects and Symbols in Art.* New York: Harper & Row, 1974.

Haller, Wilfried. "Pro. Dr. h.c.P. Josef Fischer S.J. Leben und Werk" ca. 1980. Unpublished in Archives Ignatiushaus Munich.

Harris, Elizabeth. "The Waldseemüller World Map: A Typographic Appraisal." *Imago Mundi* 37 (1985).

Harrisse, Henry. *Americus Vespuccius: A Critical and Documentary Review, Two Recent English Books concerning That Navigator.* London, 1895.

———. *The Discovery of North America.* Amsterdam: N. Israel, 1961.

Heawood, Edward. "A Hitherto Unknown World Map of A.D. 1506." *Geographical Journal* LXII (1923): 279–93.

Herbermann, Charles George. *The Cosmographiae Introductio of Martin Waldseemüller in Facsimile.* New York: United States Catholic Historical Society, 1907.

Hessler, John. "Warping Waldseemüller: A Phenomenological and Computational Study of the 1507 World Map." *Cartographica* 41 (2006): 101–13.

Humboldt, Alexander von. *Examen critique de l'histoire de la géographie du Nouveau Continent et des progrès de l'astonomie aux quinzième et seizième siecles: Five Volumes.* Paris, 1836–39.

Irving, Washington. *The Life and Voyages of Christopher Columbus.* New York: Thomas Y. Crowell & Co., 1848.

John Carter Brown Library Collections, Providence, RI.

Johnson, Christine R. "Renaissance German Cartographers and the Naming of America." *Past & Present* 191, no. 1 (Oxford, 2006): 3–43.

Jones, Edward P. *The Known World.* New York: HarperCollins, 2003.

Karpinski, Louis C. "The First Map with the Name America." *Geographical Review* 20 (1930): 664–68.

Kellner, L. *Alexander von Humboldt.* London: Oxford University Press, 1963.

King, Ross. *Michelangelo and the Pope's Ceiling.* New York: Walker & Co., 2003.

Krewson, Margrit B. *The German Collections of the Library of Congress: Chronological Development.* Washington, DC: Library of Congress, 1994.

"The Lord of the Manor and His Inheritance: An Interview with Johannes Count Waldburg." *Baden Württemberg* 4 (1997).

Magnaghi, Alberto. *Amerigo Vespucci. Studio critico con speciale riguardo ad una nuova valutazione delle fonti e con documenti inediti tratti dal Codice Vaglienti (Ricciardiano 1910).* Rome, 1926.

———. "Ancora a propisito di una nuova supposta lettera di Amerigo Vespucci sopra un suo terzo viaggio." *Bollettino della R. Società Geografica Italiana* III, ser. VII (1938): 685–703.

———. "Una supposta lettera inedita di Amerigo Vespucci sopra il suo terzo viaggio." *Bollettino della R. Società Geografica Italiana* II, ser. VII (1937): 589–632.

Malone, Dumas, ed. *Dictionary of American Biography.* Vol. VII. New York: Charles Scribner's Sons, 1962.

Matthews, Jay. "American Journal." *Washington Post,* October 10, 1988, p. A3.

Menzies, Gavin. *1421: The Year China Discovered America.* New York: HarperCollins, 2003.

Morison, Samuel Eliot. *Admiral of the Ocean Sea.* Boston: Little, Brown, 1942.

———. *The European Discovery of America: The Northern Voyages, A.D. 600–1600.* New York: Oxford University Press, 1971.

———. *The European Discovery of America: The Southern Voyages, A.D. 1492–1616.* New York: Oxford University Press, 1974.

Navarrete, D. Martín Fernández de. *Collección de los viages y descubrumientos que hicieron por mar los Españoles desde fines del siglo XV, con varios documentos inéditos concernientes a la historia de la marina castellana y de los establecimientos españoles de Indias,* 3 Volumes. Madrid: 1825–1837.

Nebenzahl, Kenneth. *Rand McNally Atlas of Columbus and the Great Discoveries.* Chicago: Rand McNally & Co., 1990.

Northrup, George Tyler. *Mundus Novus, Letter to Lorenzo Pietro di Medici.* Translated by George Tyler Northrup. Princeton, NJ, 1916.

Nunn, George Emra. *The Mappemonde of Juan de la Cosa; A Critical Investigation of Its Date.* Jenkintown, PA: George H. Beans Library, 1934.

Oftelie, Brad. "The James Ford Bell Library, University of Minnesota" *Libraries & Culture* 34, no. 4 (1999): 400–401.

Perillo, Lucia. *The Oldest Map with the Name America.* New York: Random House, 1999.

Pohl, Frederick J. *Amerigo Vespucci: Pilot Major.* New York: Columbia University Press, 1944.

Ristow, Walter W., compiler. *A la Carte: Selected Papers on Maps and Atlases.* Washington, DC: Library of Congress, 1972.

Ronsin, Albert. *Découverte et baptême de l'Amerique.* 2nd ed. Paris: Editions de l'est, 1992.

———. *La Fortune d'un nom: America, Le baptême du Noveau Monde à Saint-Dié-es-Vosges.* Grenoble: Jerome Millon, 1991.

———. *Les Vosgiens Célèbres.* France: Gérard Louis Vagney, 1990.

Sabin, Joseph, Wilberforce Eames, and R. W. G. Vail. *Bibliotheca Americana: A Dictionary of Books relating to America Vol. XXVII.* New York, 1936.

Schmidt, C. "Mathias Ringmann-Philesius, Humaniste alsacien et lorrain." *Mémoires de la Societé d'Archéologie lorraine* 3 (1875): 227.

Schnell Kunstführer, no. 1733 (1962). Munich and Zurich: Schnell & Steiner, 1983.

Schwartz, Seymour I. "The Greatest Misnomer on Planet Earth." *Proceedings of the American Philosophical Society* 146 (2002): 264–81.

Schwartz, Seymour I., and Ralph E. Ehrenberg. *The Mapping of America.* New York: Harry N. Abrams, Inc., 1980.

Seaver, Kirsten A. *Maps, Myths, and Men: The Story of the Vinland Map.* Stanford, CA: Stanford University Press, 2004.

Skelton, R. A. *Claudius Ptolemaeus Geographia Strassburg 1513 Theatrum Orbis Terrarum Ltd.* Bibliographical note in facsimile edition. Amsterdam, 1996.

Skelton, R., Thomas Marston, and George Painter. *The Vinland Map and Tartar Relation.* New Haven, CT: Yale University Press, 1965.

Small, Herbert. *The Library of Congress: Its Architecture and Decoration.* New York: Norton, 1982.

Soulsby, Basil H. "The First Map Containing the Name America." *Geographical Journal* 19 (1902): 201–208.

Stevens, Henry Newton. *The First Delineation of the New World and the First Use of the Name America on a Printed Map.* London: Henry Stevens, Son & Stiles, 1928.

———. *Recollections of James Lenox and the Formation of His Library.* Revised and elucidated by Victor Hugo Palsits. New York: New York Public Library, 1951.

Stevenson, Edward Luther. *Marine World Chart of Nicolo de Canerio Januensis.* New York, 1908.

———. "Terrestrial and Celestial Globes: Their History and Construction." *Hispanic Society of America* 1 (1921).

Thatcher, John Boyd. *The Continent of America: Its Discovery and Its Baptism.* New York: William Evarts Benjamin, 1896.

Tooley, R.V. *Maps and Map-Makers.* New York: Bonanza Books, 1961.

von Wieser, Franz R. "Die älteste Karte mit dem Namen 'America' a. d. J. 1507 und die Carta Marina a. d. J. 1516 des Martin Waldseemüller" [The Oldest Map with the Name America of the Year 1507 and the Carta Marina of 1516 of Martin Waldseemüller]. *Petermann's Mittheilungen* 47 (1901): 271–75.

———. *Die Carte des Bartolomeo Columbo über die vierte Reise des Admirals.* Innsbruck, 1893.

Warner, Deborah J. *The Sky Explored: Celestial Cartography, 1500–1800.* New York: A. R. Liss, 1979.

William L. Clements Library Files. Ann Arbor: University of Michigan.

Williamson, James A. *The Cabot Voyages and Bristol Discovery under Henry VII. With the Cartography of the Voyages by R. A. Skelton.* Cambridge, England: Hakluyt Society, 1962.

Wolfegg Castle Files.

Wolff, Edwin. *Rosenbach: A Biography.* New York: World Publishing, 1960.

Wolff, Hans, ed. *America: Early Maps of the New World.* Munich: Prestel, 1992.

Wolter, John A., Andrew M. Modelski, Richard W. Stephenson, and David K. Carrington. "A Brief History of the Library of Congress Geography and Map Division, 1897–1978." In *The Map Librarian in the Modern World: Essays in Honor of Walter Ristow,* edited by Helen Wallis and Lothar Zögner. New York: K. G. Saur, 1979.

Zweig, Stefan. *Amerigo: A Comedy of Errors.* New York: Viking Press, 1942.

INDEX

Waldseemüller world map of 1507, courtesy of Library of Congress, Washington, DC.